Magical Moons
& Seasonal Circles

Stop — Look — Listen

Stepping Into the
Circle of the Seasons

BY

SUSAN BETZ

Note to the reader: This book is an educational and informational guide. Every effort has been made to make this guide to the seasons as complete and accurate as possible. However, there may be mistakes, both typographical and in content. The remedies, historical approaches and techniques described herein are not meant to be a substitute for professional medical care or treatment. They should not be used to treat a serious ailment without prior consultation with a qualified health care professional. Some plants may cause allergic reactions in susceptible individuals. Children should use this book under the guidance and supervision of an adult.

Susan Betz has worked in the field of community gardening for over thirty years. She enjoys growing, using, studying and writing about native plants and herbs. Susan is a charter member and past president of The Michigan Herb Associates and currently writes the Children's Corner for *The Michigan Herb Journal*. She is an active member of The Herb Society of America, Garden Writers Association and the Michigan Master Gardener Association. Susan lives in Jonesville, Michigan.

Book design by Bea Osborn. Cover illustration by Jerie Artz.

DEDICATED TO

Kevin John Rooney

Is the exploration of the natural world just a pleasant way to pass the golden hours of childhood, or is there something deeper? I am sure that there is something much deeper, something lasting and significant.

There is symbolic as well as actual beauty in the migration of the birds, the ebb and flow of the tides, the folded bud ready for the spring. There is something infinitely healing in the repeated refrains of nature — the assurance that dawn comes after night, and the spring, the winter.

– Rachel Carson, *The Sense of Wonder*

Acknowledgments

A special bouquet of thanks to Jerie Artz, for painting the beautiful watercolor that graces the cover of this book. Loving thanks and appreciation to Ruth Armstrong for the interpretive black and white sketches depicting the Native American full moon names. I would also like thank my fellow members of **The Herb Society of America** for their generous support, assistance and encouragement, especially Theresa Mieseler, Susan Belsinger and Georgia Ann Mackinder.

TABLE OF CONTENTS

Introduction

Several years ago I was in the process of collecting ideas and information for use in planning and establishing a children's garden at Slayton Arboretum on the campus of Hillsdale College. I followed the advice of Jane L. Taylor, former curator and founder of the 4-H Children's Garden at Michigan State University. I invited children from local elementary schools to submit drawings depicting their visions of what gardens created especially for children should look like.

It was wise advice. The children expressed some very specific ideas on what they thought garden spaces for children should include, such as rocks, trees, water features, homes for wildlife, winding paths and resting places. Many of the drawings expressed concern for the environment by the "do not litter" signs. One drawing in particular caught my attention. It was done by an insightful fourth grade boy, named Travis, who thought gardens created especially for children should include:

- A good tree for climbing
- Strange lights for night time
- Signs that teach kids things they don't learn in school
- Big rocks
- A maze and obstacle course with flowers as a guide

I thought to myself that this young man is really in tune with nature. He is exactly the type of future gardener I was hoping to cultivate with my education and outreach programs. His suggestions reminded me that children used to gain much of their firsthand knowledge of the natural world spending time hanging out in their backyards and neighborhoods. Today many children's first introduction to gardening and nature study is in a structured class or after school program.

Mother Nature is active or impartial, mysterious, orderly, rhythmic, messy and always near at hand. She does not want you to wait for an invitation or schedule an appointment — she is ready and willing to entertain on a moment's notice. Parents, grandparents and other family relatives can help nurture a child's connection with the outdoor world by showing them that nature is not "somewhere else" but a dynamic presence in their daily lives. Following the life cycles of common local plants and animals and exploring how they respond to the changing seasons is a great place to begin. All you need do is stop, look, listen and step into the circle of the seasons.

Phenology is the study of the timing of natural periodic events in the plant and animal world influenced by the local environment, especially weather, temperature, seasonal change and climate. Examples include the first dates of budding and blooming flowers, insects hatching, bird migration, fall color and freezing or thawing lakes and ponds. This fun integrative environmental science is gaining visibility and popularity among scientists, gardeners and nature lovers across the world. Observing seasonal events and making connections help people to better appreciate and understand the importance of biological diversity. Anyone, regardless of age or educational background, can observe and enjoy the natural, cyclical occurrences unfolding daily around their backyards and local communities.

References and observations associated with phenological events date back thousands of years. Historical records and journals of past seasonal events can be used to help predict future events. By tracking changes in the cycles and times of seasonal events, scientists are better able to understand climate change and its effects on the natural cycles of ecosystems. Gardeners keep journals recording planting dates, the arrival of unwanted weeds, emerging insects, harvest times and frost dates. Nature and outdoor lovers track migrating populations of birds, butterflies and the sequence of blooming wildflowers from early spring through late fall.

In pre-colonial America, the cycles of nature directed the American Indians' daily tasks, economy and celebrations. Their knowledge of *"nature's calendar"* helped ensure their survival and kept them in harmony with the natural world. For American Indians, phenology was not only a well-honed tool, it was plain common sense. They were experts at reading their local landscapes. Life revolved around seasonal cycles and patterns of movement as they moved from one food source to another, continually modifying their behavior in response to the life cycles of local plants and animals. The emerging earthworm in March signaled the earth was beginning to thaw and the time to tap maple trees was not far off. The wild rose *(Rosa* spp.*)* blooming in June alerted them it was time to collect cedar roots and basket grass. Instead of measuring time by the weeks or months, the Indians would say a *"a moon,"* the length of time from one new moon to another. They assigned descriptive names to each full moon, alluding to notable recurring natural events taking place during that month. Variations can be found in these names across North America due to climate and temperature conditions affecting local regions. The timing of life cycle events is a fundamental ecological process. This was one of the first lessons they taught their children. All creatures great and small were created to hold a place and fulfill a purpose in the circle of life.

Magical Moons and Seasonal Circles follows the cycles of the year according to the Algonquian full moon names. The book emphasizes the effects seasonal weather and climate have on life cycles and interactions between common native plants, trees, animals and insects living in our backyards and local communities.

The full moon names assigned to each month provide the themes for seasonal outdoor activities, gardening projects, games and craft projects. Links to phenological gardening, nature study and weather Web sites are included to encourage more in-depth investigations. The book is divided by seasons and each season is introduced by an American Indian folktale.

Students, garden clubs, 4-H members, families and individuals can observe and record regional data concerning specific seasonal events and then contribute valuable information to one of numerous online citizen science projects collecting phenological data.

Phenology can be used to enhance and expand natural history, ecology lessons and gardening activities. Predictions involving folk forecasts and plant proverbs are fun to investigate and prove, true or false. Phenological studies can be pursued in formal and informal educational settings.

We, as adults, have that special body of knowledge about the natural world that we learned on our own as children — how a wild strawberry tastes, the smell of wet soil in the spring, or what happens to an apple after it falls to the ground. We, as adults, have the tremendous responsibility of providing these opportunities, along with the companionship and guidance to stimulate curiosity and encourage children down a path of self-discovery. It is my hope that adults and children will find this book an inspiring and fun guide to the seasons.

The question is not what you look at but what do you see?
— Henry David Thoreau

Tips and Tools

One of the wonderful things about exploring nature or growing plants in the home environment is that you have the time and luxury to do it the way that best fits the learning style of your family.

Having some basic tools can greatly enhance the hands-on experience for everyone and is an easy way to increase your child's observation skills and connections with the world around them. Begin with the tools you have readily available — your five senses.

Also useful:

- Hand lens
- Magnifying glass
- Camera
- Journal
- Clipboard
- Binoculars
- Pens and pencils
- Basic wildflower and nature guides

Spring

THERE COMES A DAY …

There comes a day in winter when the air

Is damp, and footprints in the snow show green—

A withered green, mocking life that's there

Beneath the sleeping snow. The crisp, blue sheen

Of ice upon the pool, catching the thin

And frosty sun, crumbles to tarnished gray,

And skeletons of leaves long dead begin

The soil's slow re-creation, with decay.

On such a day there is no hint of spring

No stirring of the earth, her pulse is low,

Dead stalks stand chill, bare shrubs their branches fling

In brittle arabesques against the snow.

Yet on this day my spirit soars the sky,

Grasping spring's glory with a secret eye.

– Ruth Matson

Why The Robin Brings The Spring

Once there was an Indian boy, the son of a noted chief. The boy had been trained to the chase and the trails of the warpath. He led in all games; he was the swiftest runner and could throw the arrow farther than any of the boys. He knew the forests and the streams, and had taught the wild game to know him. He could imitate the calls of the birds and they would flock around him. If he wandered late in the forest he had no fear of the prowling animals; he was as glad to meet the bear or the wolf as his friends in his father's lodge, for they seemed to know him and passed silently by. They boy was the pride of the village and the boast of his father who believed that he would be a great chief.

The time of his dream fast came. He must fast at least seven days and his clan spirit must appear to him in the form of a bird, animal, reptile, fish, a tree, a plant or a root. The snows were deep and the winds keen, but the boy was young and his blood like fire, and he welcomed the fast. To endure, that was his birthright and his boast.

He built his lodge out of young saplings in the heart of the woods. He covered it with branches of green hemlock to shelter him from the snows, and taking off his furs and appealing to the spirit of his clan, he went in.

His fast had begun and he was alone with his thoughts. He had been kind and happy. No frown had come to his lips, nor sorrow, and now his manhood was approaching. Ten suns to pass above him; ten nights for his dream. If the deer, he would wind its soft skin about him to warn away the cold winds. If the bear, he would string its strong claws to wear around his neck. If the wolf, his white teeth would guard him from danger. If the turtle, his shell would be a breastplate. If a bird, his wings would adorn him. He had no thought but faith in his dreaming.

Nine days lighted the forest; nine days darkened the lodge. The tenth day dawned frowning and gloomy and the chiefs came. They shook the lodge pole and bade him appear.

"Not yet to-day," said the boy. "I have to see the clan spirit three times, and, although I have fasted and prayed, the clan spirit has come but once. Return to-morrow."

Again on the morrow the chiefs came. "One day more," pleaded the boy, although his voice was very weak, for if the dream did not come the chiefs would release him and he would depart unhappy and in disgrace.

Again on the morrow the chiefs came. They announced that the time was past, and again he implored for one day more.

"If the spirit doesn't come by to-morrow," he said, "I'll go. To-morrow I will depart with you." His voice was weak and the chiefs were anxious. Cautiously they parted the hemlock branches and they saw the boy panting.

On the morrow the chiefs again shook the lodge pole. There was no response. A strange silence had fallen over the forest. The awed chiefs wondered and entered the lodge. The boy was not there, but a bird flew down to the branch over the lodge and began to speak.

"I am he whom ye seek. My body is no more on earth. I fasted and waited, but no dream came. I didn't know the reason. I had done no evil. Death was the friend who aided me to flee the disgrace, which would follow me if I were denied my dream. Now I am the Robin."

"Do not sorrow, nor mourn me. I will bring the spring to you. I will sing to the trees and young leaves will come forth to listen. I will swing on the wild cherry and its blossoms will welcome me. I will carry the gray shadows of the spring morning on my wings. I will not hide in the forest. I will nest by your lodges. Your children will know the spring is coming when they hear my voice. Though the snowfall may cover my path, it will melt into singing streams when it hears my wings rustling."

"I was willing, and I painted my body red when I felt my spirit departing and know I carry its red glow in my breast as its shield."

"He was brave," said a chief. "We should have taken him sooner."

"Now," said his father proudly, "he is the Robin who brings us the spring."

"He is the Robin forever," chanted the birds.

"Why the Robin Brings Spring." Published in *Why-So-Stories* by Edwin Gile Rich, Small, Maynard and Company, Publishers (1920) pp. 33-38.

MARCH

March can only be considered in charity a spring month;
there is plenty of winter left, the bite and steel cold wind, frost
in the mornings, rain, a pallid wash blue sky, sunshine with
a nip in it. March is the meeting and duel of winter
and spring. Spring starts forward, moves back;
there are days of growth and non-growth.
Yet the hours of light are lengthening.

– Anita Nygaard, "Earth Clock"

Sugar Moon — Worm Moon — Crow Moon

March, the *awakening month*, displays nature's visible response to rising temperatures and increasing hours of daylight, signaling the retreat of winter and the arrival of the Sugar Moon. March ushered in one of the most anticipated seasonal events in the social and economic lives of Indians living in the Midwest and Northeastern regions of the United States.

The Sugar Moon alerted them that spring was near and it was time to return to their sugar camps to tap the maple trees and collect the sap. They would leave a lodge at the camp all year round along with a storehouse in which they kept their tools for producing maple sugar. A food cache was always left at the camp in the fall, filled with dried foods ready and waiting for them when they arrived back in the spring.

Families would work together tapping the trees, tending the fires and cooking the sap. Three kinds of sugar were made from the boiled sap. *Grain Sugar* was produced by boiling and stirring the sap till it crystallized. *Cake sugar* or thickened sap just before crystallization was poured into wooden molds shaped like animals, flowers and stars to make sugar cakes that were saved and used for gifts. The third type was *gum* or *sugar wax*, where thick strings of boiled sap were tossed on the snow and turned into taffy-like strands that children would chew, play with and eat. Maple sugar was used by American Indians throughout the year to preserve berries and season fish, vegetables, grains and game. It was mixed with water to make a maple-flavored drink. To learn more about the history of maple syrup, American Indian sugar camps and traditions, visit http://www.kstrom.net/isk/food/maple.

"Who knew that earthworms were one of our planet's most important caretakers"?
– *The Boston Globe*, 2005

A traditional sign of spring is the return of the American Robin, hopping about the lawn and garden looking for earthworm castings. A robin can eat ten to twelve worms in an hour. As the temperature warms and the ground begins to thaw, earthworms emerge at night exploring the soil surface in search of food. Collecting bits of leaves, seeds and animal remains, the worms drag them down into their burrows where they decompose and are easily digested.

Soil is ingested along with the decaying matter, and then expelled as little mounds of castings outside the opening of their burrows. Giant ragweed seed is a favorite item of earthworms to gather and bury in their burrows. It has been speculated they find the seeds an appealing structural element for strengthening their burrows while providing a future food supply. By burying the seeds they help contribute to the plant's survival and widespread dispersal.

Earthworms increase soil fertility, merely by going about their daily activities, plowing about beneath the earth's surface loosening the soil and breaking down organic matter. Their tunnels provide air and drainage space for rainwater to filter down through the soil. In dry and cold weather these burrows can extend down six or seven feet beneath the soil surface. Their presence in yards and gardens is a good thing; it is an

THE ROBIN

There came to
my window
One morning in spring,
A sweet little robin;
It came there to sing.
And, the tune
that it sang
Was prettier far
then ever
I heard on flute
or guitar.
– Dunn & Troxell
Selected Mother
Nature Series

indicator of the proper functioning and fertility of the soil. The American Indians knew the important function worms performed in the great circle of life, and thus named a full moon in their honor.

If Crows Are Seen In February, There Will Be An Early Spring

Full Crow Moon links the sound of cawing crows to the end of winter and the arrival of spring. Crows are alleged to be one of the most intelligent and clever of our native birds. Aesop's fable, *The Crow and the Pitcher,* is an ancient tale about a thirsty crow that by chance comes across a pitcher of water with a level too low from the pitcher top for him to reach and drink. He cleverly raises the water level by dropping stones into the pitcher, demonstrating *"little by little does the trick."* Researchers recently released a crow study that found rooks, members of the crow family, can indeed use stones to raise the level of water in a container, just like the clever bird in the fable written thousands of years ago.

For a brief period in the spring, crows become extremely noisy and quarrelsome with one another before settling down to nest. During the nesting season, crows scatter into small groups, thought to be composed of a breeding pair and their non-breeding offspring born the previous year. They do not gather again in flocks until the spring hatchlings are fully grown and are able to leave the nest. In the autumn, some will move southward, but most stay for the winter gathered in large communal roosts in pine or hemlock forests or other evergreens.

Spring

Traditionally, spring arrives on the night of March 21 or 22 depending on the day of the vernal or spring equinox. Each day following the winter solstice, the path of the sun travels a little higher in the southern sky, ris-

Ere Man is aware
That the Spring is here
The flowers have
found it out.

– Old Chinese saying

ing a little closer to the east and setting closer to the west, until the day when it rises exactly east and sets exactly west. This is called an equinox, when the days and nights are of equal length all over the world. Vernal originated from the Latin word for *bloom*, signifying the arrival of spring in the Northern Hemisphere, when buds of flowering trees and plants begin opening.

Spring travels north, west and uphill at predictable rates: 15 miles north, 1 degree west and 30 meters uphill a day. After the spring equinox, the sun continues to follow a higher and higher path through the sky, with the days growing warmer and longer, until on the summer solstice it reaches its highest point in the sky. You can actually track the progression of spring as it moves northward. *Journey North*, www.learner.org/jnorth/tulip, is an outstanding Web site packed with resources, information and projects for families and educators interested in following the seasons.

Many popular superstitions and prophetic warnings have been attributed to March's unsettled weather as winter turns into spring. Old folklore depicts the first three days of March as blind days because they were considered unlucky. Supposedly, if rain fell on these days farmers would have poor harvests. "*A peck of March dust and a shower in May, makes the corn green and the fields gay.*" Many farmers were so superstitious about the first three days of March they would wait to plant their seeds after these days had passed. The lamb and lion have long been associated with unpredictable March weather. "*March comes in like a lamb, it goes out like a lion, If it comes in like a lion it goes out like a lamb.*" This ancient weather proverb has its origins with the constellations Leo the Lion and Aries the Ram or Lamb and their relative positions in the sky at the beginning and ending of the month. Leo rises in the east at the beginning of March and Aries sets in the west at the end of March.

SPRING CONSTELLATIONS EVERY CHILD SHOULD KNOW

Leo, the lion
Aries, the Ram
Scorpia, the Scorpion
Ursa Major, the Great Bear
Ursa Minor, the Little Bear

BIRD TRADES

The Swallow is a mason,
And underneath
the eaves,
He builds his house
and plasters it
With mud and straw
and leaves.

The woodpecker is
hard at work,
A carpenter is he,
And you can hear him
hammering
At his house up
in the tree.

Of all the weavers
that I know,
The oriole is best,
High in the branches
of the tree
He hangs his dainty nest.

— ANONYMOUS

Backyard Bird Watching

A large number of our familiar songbirds start to migrate north in late winter and early spring. Late March is a great time to begin observing the life cycles and social behavior of the birds living in your neighborhood and yard. The male is most often the one heard singing; it is his way of attracting a mate and a musical warning to other males to stay out of his territory.

The American Robin and Common Crow are two of the first birds to start nesting in spring. Birds are just as picky and particular about their homes as people are. Robins like to live and nest around humans. Crows become very guarded and secretive, building their nests in the tops of tall pines and oak trees away from humans.

Keep a pair of binoculars and a family wildlife journal handy to observe and record the activities and habits of the different species of birds arriving to set up housekeeping and raise their families. The majority of birds are active from sunrise until 1:00 p.m. and then again from about 3:00 p.m. until sunset.

Make sketches of the birds, noting the differences in their heads, bodies and beaks. Be sure to include the trees, grass and other interesting natural features of the landscape where the birds hang out. Field guides can be used to add the finishing detail to your drawings.

By just observing carefully — no books necessary — you can answer all of these questions!

• Where do they build their nests?

• What are the nests made from and how are they constructed — loosely, tightly, woven or stacked?

• Are there specific trees or shrubs they like to build their homes in?

• How are the nests attached to the stems and branches of the trees and shrubs?

- How many eggs do they lay and do they all hatch?
- What do they eat — seeds or insects? Where do they eat — off the ground or from a feeder?
- Some birds will just be passing through, stopping to rest and refuel before continuing on northward to their summer habitats. How long do they stay before moving on?
- Which birds hop or walk — or do both?
- How do birds communicate with other birds?

Citizen Science Project: *NestWatch* is a nest monitoring project developed by the Cornell Lab of Ornithology in cooperation with the Smithsonian Migratory Bird Center and funded by the National Science Foundation. The Web site provides detailed instructions on how to safely observe and identify nesting birds. Log onto www.birdsource.org.

The Sibley Guide To Birds and *The Sibley Guide to Bird Life and Behavior*, by David Allen Sibley, and the Stokes Nature Guides, *A Guide to Bird Behavior Volume I & II,* by Donald and Lillian Stokes, are two fine field guides to help with identification of common North American birds.

Birds of Algonquin Legend, by Robert E. Nichols, Jr., is a delightful collection of stories and legends gathered from the Algonquin Indians of the Northeastern United States.

Nourishing Nature Naturally

Mother Nature knows best, so begin planning now to follow her lead and switch to nature-based methods to maintain your gardens and yard. Earth-friendly landscapes and gardens are family-friendly landscapes and gardens. March is a perfect time of year to walk around your yard and take an inventory of your landscape and gardens. Keep the following

Sweet April showers
Do spring
May flowers.
– Thomas Tusser,
A Hundred Good Points
of Good Husbandry, 1557

*Pussy willow
had a secret,
That the snowdrops
whispered to her,
And she purred it to the
South wind, while it
stroked her velvet fur.*

*And, the South wind
hummed it softly to the
busy honey bees,
And they buzzed it to
the blossoms on the
scarlet maple tree.*

*And, they dropped it to
the wood-brooks,
Brimming full of melted
snow, and the brooks
told the robin red breast,
As they babbled
to and fro.*

*Little robin could not
keep it, so he sang it
loud and clear,
To the sleepy fields
and meadows,
Wake up! Cheer up!
Spring is here!*

— Anonymous

questions and tips in mind as you stroll about your yard.

• Learn as much as you can about your soil; then you will be better able to match plants to your conditions. A simple way to determine the type of soil you have is to grab a handful and give it a squeeze. If you have sandy soil it will crumble and won't hold its shape in your hand. If you have clay soil, it will form a sticky lump when you squeeze it. Loam, the ideal garden soil, will form a ball that breaks easily when you squeeze it. The health of all soils can be improved by the addition of compost.

• Do you have any microclimates? Microclimates are small pockets in your landscape that differ from the dominant landscape in temperature, sun exposure and moisture.

• Which areas retain frost and collect snow?

• Trees and shrubs vary considerably in their ability to withstand exposure to wind. Some are able to withstand severe exposure and still thrive, while others require some protection from winter wind and sun. Winter winds blow from the northwest and summer winds come from the southwest.

• How many hours of sunlight does each portion of the landscape receive? Full sun equals more than six hours of direct sunlight a day; a half-day of sunlight equals four to six hours each day; shade equals two to four hours a day and dense shade equals zero to two hours a day.

The basic idea behind naturalistic gardening is that well-chosen plants on a lean but adequate diet are healthier and more sustainable than artificially pampered plants in excessively lush conditions. By reducing the hard surfaces and increasing green space you can create landscapes that blend with and support the natural environment, providing a welcome retreat for both humans and wildlife.

State and local water restrictions are becoming increasingly commonplace in every part of the country. Xeriscaping is the norm in the Western regions of the United States. Landscapes are designed using plants with water requirements corresponding to the typical local rainfall patterns. Mesiscaping, landscaping to reduce water usage in mesic regions (moderately moist), has become an equally important practice for people living in the Midwest and Northeastern regions of the United States. A local nursery can help you choose trees and plants that match with your soil, family needs, climate and seasonal weather patterns. *Grow Native*, www.grownative.org, provides free landscaping guides showing you how to create earth-friendly yards and gardens.

Composting has many advantages which are beneficial to the life cycle of all living things. It is a simple way to improve your soil texture, build fertility and increase water retention capabilities. Some people just dig holes and bury their vegetable scraps in their garden beds and borders; others use compost bins. To find which method is most appropriate for your lifestyle visit www.mastercomposter.com.

Vermicomposting is the practice of using worms living in bins to process organic household waste into nutrient-rich compost. Scientists classify worms by the function they fill in the ecosystem. Red-wigglers *(Eisenia fetida)* are used for vermicomposting because they can eat half their weight in a day. Two thousand worms will eat 1 kilogram or 2.2 pounds of vegetable waste a day. These are a different species of worm than the earthworms commonly found in the yard and garden.

Red-wigglers like to live close to the soil surface, thrive in confined spaces, and do not mind being disturbed by the lid of a bin being lifted to add vegetable waste. A worm composting system is not hard to maintain and is an interesting way to learn about the functions of worms in our ecosystem while turning kitchen waste onto valuable compost. The castings they produce are useful to add to potting soil to help promote healthy plant growth. Night crawlers and other garden worms prefer to

burrow deeply in the soil, do not like to be disturbed and are not able to process large amounts of organic matter. Do not use these worms for vermicomposting. For more information and help getting started with vermicomposting, composting and organic soil management, check out www.wormwoman.com. On this site you'll find Mary Appelhof's classic, *Worms Eat My Garbage*, along with worm bins and worms for sale.

Watch the robins in your yard and neighborhood; they can direct you where to find the worm castings through spring and early summer. Charles Darwin spent years studying the behavior of earthworms; he admired their work ethic and was delighted by their intelligence. He discovered they were smart enough to collect all sorts of things, including tiny paper triangles he placed around their burrows. After you locate some castings, try placing some tiny paper triangles around an earthworm burrow opening in late evening and then check it out in the morning to see what is left. Look in the burrow to see if the worms drag them down inside their tunnels. For more fun activities with worms check out http://kids.niehs.nih.gov/worms.

Skunk Cabbage (*Symplocarpus foetidus*)

The skunk cabbage is one of the oldest known wildflowers in North America. It has many folk names — swamp cabbage, parson in the pillory, polecat weed, Midas ears and hermit of the bog. American Indians used the plant medicinally and made a soap infusion of the dried and powdered root as an underarm wash to eliminate body odor. Landscapers and surveyors depend on the skunk cabbage to indicate damp ground and wetland areas.

As the ground starts to thaw and the snow and ice begin to melt, thick brownish-green, striped, pointed hoods can be seen poking out of the mud in marshes and swampy areas. If you look closely inside at the bot-

Skunk Cabbage

tom of the thick little hood, you will discover a short, stout stem covered with tiny yellow flowers dusted with yellow pollen. Skunk cabbage flowers are one of the first food sources for early emerging pollinating insects. Temperatures within the little hood can heat up to 70°F, providing a perfect place for tiny flowers to live and host visitors during the cold days of early spring. Another interesting characteristic of the first flower of spring is its distinctive odor — a mingled scent of garlic, skunk and putrid meat. Tiny black flies, one of the first insects to emerge in the spring, find the plant a delectable place to gather and dine on stinky pollen. Carrion beetles are also attracted to the foul smell and can be found crawling in and about the base of the plant.

Honeybees can sometimes be found hovering beneath the warm hood in early spring before their other food sources are available. After the plant is pollinated, the leaves begin to unfold, growing up to two feet across by early summer. During the summer, frogs and lizards make their homes beneath the large green leaves. The Common Yellowthroat bird sometimes builds her nest in the hollow of the large cabbage-like leaves. Then by late summer the plant dies back to its roots and disappears to wait for spring.

March is the perfect time to scout for praying mantis egg cases in the weeds and shrubs. They are most often found at eye level. They look like foamy, tan-colored marshmallows attached to leafless shrubs, weeds or rocks and buildings. There can be up to four hundred eggs inside each case waiting to emerge on a warm, late April or early May afternoon. The praying mantis is a beneficial insect that feeds on a variety of insects they catch with their specially adapted front legs. It is the only insect that can look over its shoulder. If you can't find an egg case in the wild, check some of your local nurseries or buy one online. Watch the weather temperature as the days warm; if you are lucky you can catch them emerging from

Praying Mantis

Praying Mantis Egg Case

the egg case. Keep a careful eye out during the summer. The praying mantis are masters at camouflage so they can easily be missed perching among the vegetation in your yard and garden. They are very fond of scented geraniums and are fun to keep as pets throughout spring and summer. They become quite large by the end of summer, two to four inches long. After mating in late summer the females begin laying eggs in a frothy liquid that turns into a hard protective shell.

To learn more about beneficial insects and how to identify them obtain a copy of *Good Bugs For Your Garden,* by Allison Mia Starche. It is written like a field guide with helpful descriptions of the insects and the jobs they perform.

Mad as a March Hare

March is the breeding season for rabbits and hares. The mad March hare from *Alice's Adventures in Wonderland* was based on the old saying, *Mad as a March Hare.* The males dance and dash about sparring with each other to establish dominance and secure access to breeding females. It was long assumed that the demonstration was limited to inter-male competition, but females have also been observed "boxing" males with their paws, reminding them to mind their manners and state their intentions in a more respectful manner.

Rabbits and hares are related but are different species. All rabbits, except the cottontail, live in colonies in underground burrows. Baby rabbits are born blind and naked and unable to care for themselves until about three weeks of age. Hares live above ground in simple, untidy nests and as adults rarely live together socially. Baby hares are born with vision and fur and are left to fend for themselves a few days after they are born.

Rabbits and hares are associated with the moon in many cultures around the world; figures of white rabbits or hares are common at Chinese Moon Festivals. Carvings of rabbits appear on ancient Greek and Roman tombs, symbolizing the transformative cycle of life, death and rebirth.

Peter Rabbit's misadventures in Mr. McGregor's vegetable garden is an all time favorite children's story with a gardening theme. Planning and planting a theme garden based on the beloved books by Beatrix Potter is a wonderful way to introduce young children to gardening. Michigan State University Children's Garden has a Peter Rabbit theme garden that you can tour online for plant suggestions and layout. The MSU Children's Garden has fifty-six different themed gardens. Browse their Web site at www.4hgarden.msu.edu/tour/overview. You will find a wealth of creative ideas to help design a fun family or community garden.

GOOD FRIENDS — OUR NATIVE TREES

The Sugar Maple *(Acer saccharum)*

There are over a hundred species of the maple family growing throughout the Northern temperate regions of the world. Thirteen of those are native to the United States. The Sugar Maple is one of our largest and best loved deciduous trees. It has been valued for its beauty and utility since ancient times.

Maple sugar was one of the most important staples in the domestic economy of American Indians living in pre-colonial Northeastern regions of the United States. Maple sugar was used in much the same manner early Europeans used salt to preserve and flavor their food. They showed French and English explorers who came to America how to tap the tree for its sweet water and make maple sugar. The hard wood was prized for making tools and canoe paddles. Maple trees were associated with success and abundance, and the maple leaf was an important element in

The Maple puts her corals on in May, While, loitering frosts about lowlands cling, To be in tune with what the robins sing, Plastering new log huts 'mid her branches gray: But when the Autumn southward turns away, Then in her veins burns most the blood of Spring, And every leaf, intensely blossoming, Makes the year's sunset pale the set of day.

– JAMES RUSSELL LOWELL

American Indian beadwork designs. It still ranks as one of America's most important hardwood trees for making furniture, paneling, flooring and musical instruments.

Maple trees have inconspicuous clusters of greenish-yellow flowers that appear in late March, blooming before the trees leaf out. Because of its discreet flowers, the maple tree was considered an emblem of reserve in ancient cultures. Maple trees are primarily pollinated by the wind, with occasional assistance from our native bees and other early emerging insects. Once pollinated, the flowers produce interesting paired seeds with wing-like membranes, called keys, which help distribute the seeds on the wind as they fall from the trees. Children used to make wishes on maple keys found on the ground and it was considered good luck to catch a spinning key in your hand as it fell from the tree. During the active growing season, maple leaves can be identified by their hand-like shape growing on opposite sides of the twigs.

The trees can reach heights of a hundred feet or more and thrive in a variety of locations, with the exception of sandy or poorly drained soils. During the summer its leaves provide a cool shady cover for people and wildlife. The Sugar Maple is often described as the king of the shade trees and is legendary for its glorious fall color, when the leaves change to brilliant shades of crimson, orange and yellow.

Maple trees support a wide variety of forest dwelling butterflies and moths. Moose, deer, rabbits, squirrels and porcupines and many species of birds feed on maple bark, buds and twigs. Pileated woodpeckers and screech owls like to nest in maple trees.

ACTIVITIES AND ADVENTURES FROM
THE BACKYARD AND BEYOND

PUSSYCATS ON A FENCE

Materials needed:

- Sturdy paper
- Colored pencils
- Dried flower stems and leaves
- Catkin flower clusters
- Tacky glue
- Pussy willows
- Black pen

Draw and color a backyard scene, then build a picket fence using dried brown plant stems for the slats of the fence. Glue the pussy willows along the fence. Using a black drawing pen, add some ears. The longer droopy catkins from the aspens and beeches make perfect tails. Use a hand lens to compare the different shapes and sizes of the catkin flower clusters.

WILLOW WHISTLE

Early spring, when the sap is running, is the best time to whittle a willow whistle. Pick out a willow branch roughly 4 – 6 inches in length and about 3/4-inch in diameter with smooth bark and no side shoots sticking out. To make a mouthpiece, cut off one end at an angle and square off the point. Turn the willow piece over and cut a notch through the bark about 3/4 of an inch from the mouthpiece.

Make a circle and cut just through the bark about three inches from the mouthpiece. Gently tap or rub a bowl of a spoon until juice runs and the bark loosens all the way around. Be very careful not to crack it; the bark should slide off in one piece.

Carefully remove the bark cylinder (mouthpiece) and set aside. Lengthen and cut the notch deeper in the wood and cut a small sliver off the angled end. Slip the bark cylinder back onto the willow stick and blow.

Beaufort Wind Scale

Calm

Gentle Breeze

Fresh Breeze

Moderate
Gale

Violent Storm

<small>Watercolor courtesy Jeri Artz</small>

The Beaufort Wind Scale
was devised by
British Rear Admiral
Sir Frances Beaufort in
1805. It was based
on observations of
the effects of the
wind at sea.

MAPLE SEED DRAGON FLIES

Materials needed:

- Maple seeds
- Elmer's glue
- Tacky glue
- Glitter
- Twigs an inch or so long
- Tiny beads

Paint the maple seeds with a light coating of Elmer's glue, shake on the glitter and let dry. Glue two pairs of maple seeds to the twigs for wings and glue beads in place for the eyes.

BACKYARD WEATHER STATION

A backyard weather station is relatively easy to assemble and will provide great family fun. It will supply practical clues on how to dress for daily weather conditions and increase children's observation skills while satisfying their natural curiosity about weather. A simple weather station is a collection of devices for measuring natural elements such as temperature, rainfall, wind speed and atmospheric pressure. Locate areas in your yard to mount your weather implements that are easy to access.

- Thermometer – to record temperature
- Rain Gauge – to measure rainfall
- Barometer – to measure atmospheric pressure
- Anemometer – to measure wind speed
- Wind Sock – to determine wind direction

THYMELY TIPS AND SAGE ADVICE

March 14 is National Pi-Day

- Whatever the size of a circle, if you divide its circumference by its diameter you will always get PI= 3.14159. The circle is a basic shape found over and over again in the natural world. The word circle comes from the Latin word *circa*, which means day. A day is twenty-four hours long. Twenty-four hours is the time it takes the Earth to spin completely around once on its axis. A full day (one day-night cycle) or circa is a type of a circle. Keep a list of all the different circles you see around you in a day.

• Apart from earthworms, a few of the other insects living below the soil surface are mites, ants, spiders, millipedes, beetles, sow bugs and springtails. It is pretty amazing when you consider that there are more living creatures per cubic foot living in the community beneath our feet than there are people on earth. *The Earth Moved, On the Remarkable Achievements of Earthworms*, by Amy Steward, is a fascinating book that explores the subterranean world beneath our feet. She does a wonderful job describing the *"soil food web"*, that incredibly complex network of soil critters from earthworms down to the smallest bacteria living and working together. After reading this book you will understand the American Indian's respect for these lowly organized creatures and their importance to the environment

• A simple rule to remember when transplanting trees and shrubs. "Evergreens may be transplanted from the time the leaves of the lilac are the size of a mouse's ear until the leaves fall off in the fall. Deciduous trees just the opposite, from the time the leaves fall in the autumn until the leaves of the lilac are..." Of course this does not mean that it is a good idea to transplant your rhododendrons at the end of July or try to hack out a young oak in frozen January. It's merely a gentle reminder of general planting times for woody plants.

– *The Home Garden Magazine,* February, 1946

• Now is time of year to get a soil test. The pH range where nutrients are readily available to plants is 6.5 to 7.0. You can buy a soil kit at your local extension office. The extension staff members can also help you

Circles are one of the most common shapes found in nature. Pictured, water dripping on a rock and creating a circle.

read and understand your test results.

• According to ancient British folklore, to rid your house of fleas you must burn a dirty dish cloth when the first thunder cracks in March. If that does not work you can try sprinkling your house with dirt scooped from beneath your right foot when you hear the first cuckoo call of the year.

SONG OF THE WIND

I am a giant strong and bold.
Such jokes I play on young and old;
But I work hard from sun to sun,
And one must have a little fun!
Sometimes a boy I chance to meet,
I blow his hat across the street,
Then toss the kite up to the sky.
And help his mother's clothes to dry.
The flags I wave, the pin-wheels turn,
The blacksmith's fire I help to burn.
Then when it rains I frisk about.
And turn umbrellas inside out.
I send down leaves in golden showers.
To make warm blankets for the flowers.
And then again the seeds I sow,
I bring the showers to make them grow.
And then I go far out to sea.
Where many boats still wait for me.
And when the evening sky is red
I take the fishermen home to bed.

– M. Helen Beckwith

APRIL

"Sing a Song of Spring" cried the pleasant April rain

With a thousand sparkling tones upon the window pane,

And the flowers hidden in the ground

woke up dreamily and stirred,

From root to root, from seed to seed,

crept swiftly the happy word.

– CELIA THAXTER

Sprouting Grass Moon — Frog Moon — Egg Moon

Spring is a transitional season that comes only to the temperate zones of the world. In the Northeastern regions of the United States, the appearance of the natural world changes more dramatically during April than any other month of the year. The last of the snow and ice disappears, the trees and shrubs start to show off their leaves and grass begins to green and grow. Small animals that commonly sleep through the winter begin emerging from their burrows and homes beneath rocks and rotting logs. A few early bees and butterflies venture out seeking nectar from the first flowers of spring.

For the Algonquin people, fresh green grass sprouting from the earth was proof that spring had indeed arrived, increasing anticipation of the earth's fruitful abundance; hence the name Sprouting Grass Moon.

Grass is one of the largest and wide-ranging families in the plant kingdom, surpassed only by the orchid, legume and daisy families. The fruits of grass plants are called grains or cereals. Rice, wheat, rye, barley, corn and oats have been an essential food source for humans and animals for thousands of years. Sugar cane is a type of grass that provides more than half the world's sugar supply.

The stems and leaves of some grasses are used to make paper. Grazing and forage grasses provide most of the feed animals eat; turf grasses are used to cover golf courses, playgrounds and athletic fields. There are roughly 40 million acres of lawn in the United States, making it one of the largest irrigated crops in the country. It surpasses even corn (maize) acreage. Irrigation of lawns accounts for 40 to 80 percent of water used in city and suburban residential areas, and this is an area of intense ecological debate. Is it worth the effort? Ornamental grasses are used in flower beds, parks and other landscaped areas because of their beauty, form and low maintenance. Other types of prairie grass and native plant-

GRASS WHISTLES

In bygone days, children used blades of grass to make loud, whistling noises. Find a wide blade of grass four to five inches long. Hold it parallel between both thumbs. Flip the blade vertically and holding it pressed tightly together up to your lips, blow through the vertical cracks between the thumbs and around the taut blade of grass.

ings are being implemented in lawn designs to provide beauty and low water consumption.

Grass was used to symbolize submission and the fleeting quality of life in ancient works of poetry and art. American Indians used grasses for decorative purposes, to make dolls for their children, and for weaving baskets and mats. In the fall they would gather dried grass into bundles to line the walls of their winter cache pits. American Indian women enjoyed a game called "Bunch of Grass". They would tie a bunch of grass by a short cord to a pointed stick and try to catch the grass onto the sharp point of the stick.

The chorus of singing frogs and toads in the marshes and woodland ponds is as much a part of spring as the return of migratory birds and appearance of the first wildflowers. The American Indians knew the sound of trilling toads and singing frogs signaled the beginning of their mating season and the renewal of the earth's life cycle. In Native American folklore, the frog was often depicted as the guardian of the fresh water in springs and wetlands around the world.

Frogs and toads are classified as amphibians and make up about 2.5 percent of all living species of animals with backbones. The word "amphibian" means double life in Greek. Amphibians are cold-blooded animals that live a dual life. Most live the first part of their lives in water as aquatic larvae with gills, then transform into animals that live on land. Frogs and toads breathe through their porous skin and live in wet or damp places. Both species share a similar life cycle beginning as tadpoles but grow into amphibians. They are considered excellent ecological indicators because of their dual life cycle and sensitive permeable skin. If the frog population begins to decline in an area it is usually an indication that something in the environment is changing.

The first 50°F day will usually trigger the call of the little spring peeper,

APRIL NIGHT

From every little
wayside pool
The hylas' silver
sleigh-bells ring,
And hate and death
seem but a dream.
Lost in this lovely
night of Spring.
– Marion Doyle

a faithful harbinger of spring. An old agrarian proverb passed down through the centuries cautions *"frogs will look through ice twice,"* reminding gardeners there will be two more frosts after frogs sing in spring.

One of the most common questions people ask is *"What's the difference between frogs and toads?"*

Frogs have bulging eyes, smooth moist skin, live mostly by water and have long legs for jumping. Frogs are easy to study during their mating season because each species has its own distinctive call that will tell you where and what they are.

Toads spend almost all their adult life on land and temporarily go to water to breed in the spring. They have bumpy skin and hop along on stocky short legs. The musical trilling of the toad begins hesitantly in early April, swelling to a loud chorus by the end of the month. Toads have a special defensive mechanism that helps deter predators. If attacked they exude a nasty tasting substance that can burn the mouth and eyes of potential predators. Italian violinists in the 17th century would stroke toads to keep their hands from perspiring during their performance. The alkaloid substance the toads secrete dried out the skin on the violinist's hands.

Frogs and toads eat a variety of insects, some of which are pests to humans. They are useful friends to have around the yard and garden. One modest-sized toad can eat up to 26 insects a day between May and August. Avoid using pesticides and fertilizers in and around places they live.

Like seeds, eggs are an ancient symbol of spring, rebirth and renewal. Amphibians and other animals give birth in April, which is why some American Indians referred to April as the Full Egg Moon.

All through April, frogs, toads and salamanders are breeding and laying their eggs. Their eggs come in all shapes and sizes. The American Toad's eggs look like a string of black pearls. The Spotted Salamander's eggs are

The frog does not drink up the pond in which he lives.

– INDIAN PROVERB

Frogs are lucky – they eat what bugs them.

small, jelly-like masses attached to twigs submerged in fishless ponds and wetlands. Frogs lay their eggs in large gelatinous masses.

April is a perfect time of the year to collect and examine the eggs of frogs, toads and salamanders with a hand lens. While you are searching for the eggs of these amphibians, watch for insects hatching from eggs laid late in the summer and fall.

Water, water everywhere — April is wet and watery. The thawing ground is soggy with water and water is in the air as rain, snow, ice, steam and fog. Our planet's water has been recycled time and time again from the very beginnings of the Earth. The Earth's water cycle is nature's ultimate recycling system. No new water is being manufactured. We have the same amount of water we have always had or will have in the future. The water the dinosaurs drank millions of years ago is the same water that falls as April showers today.

Government studies have shown that as much as 70 percent of the pollution in our streams, rivers and lakes is carried there by storm water rushing off roofs, sidewalks, paved parking lots and roads. In the natural environment, the soil and plants act as a sponge, absorbing most of the water that falls to ground, creating a natural watershed. The water soaks into the soil replenishing the groundwater supplies. It is taken up and filtered by plants, then returns to our ponds, streams and lakes clean and unpolluted. Studies have shown that natural undisturbed woodlands can filter and recycle five times more rain per hour than a cleared woodland or conventional lawn before runoff occurs.

A Thoughtful Solution to Water Pollution

A rain garden is a special kind of garden designed to collect and absorb runoff from roofs, lawns or parking lots that normally rushes into sewers or local waterways. Instead, water is slowly filtered by plants and soil in

FIRST FLOWER OF SPRING

In spring, the day you find the first flower of the season can be used as an omen:

Monday means good fortune.

Tuesday means greatest attempts will be successful.

Wednesday means marriage.

Thursday means warning of small profits.

Friday means wealth.

Saturday means misfortune.

Sunday means excellent luck for weeks.

the garden. Planting a rain garden in your yard or neighborhood will help ensure the health of your local waterways. It will also provide food and shelter for wildlife and add hardy, low maintenance and naturally beautiful garden to the neighborhood.

Rain gardens can be designed for any site — shade, sun, wet or dry — and created in all different shapes and sizes. They are super easy to install and maintain by following these simple steps:

- Pick a naturally low spot in your yard located at least ten feet from your house where runoff can be delivered away from building foundations and utilities.

- Dig a shallow depression with a level bottom, as large in circumference as you would like.

- Direct your downspout or sump pump outlet to your rain garden either by digging a shallow swale or linear depression designed to channel water, or by routing it through a buried 4-inch PVC pipe.

- Species of perennial plants native to your region are the best choice for a rain garden because they are adapted to local conditions and will not need extra care once they are established. Include some native ornamental grasses and sedges as well as wildflowers to ensure the garden has a strong root mass that will resist erosion and inhibit weed growth. Keep the garden well watered for the first few weeks until the plants are established.

- Leaving the seed heads and spent foliage in place through the winter will provide cover and food for many kinds of wildlife.

This information was adapted from *Build Your Own Rain Garden, Perennial Garden Design Sheet #1*, published by Gardener's Supply Company in Burlington, Vermont. For more information on tools and water conservation resources visit their Web site at www.gardens.com.

Sweet April showers
Do spring
May flowers.
— THOMAS TUSSER,
A HUNDRED GOOD POINTS
OF GOOD HUSBANDRY, 1557

April hath put a spirit of
youth in everything.
— WILLIAM SHAKESPEARE

Take a Walk on the Wild Side

If you look around your yard and gardens, grass shoots are not the only plants you will see greening the spring landscape. Many of the plants we dismiss as weeds are European migrants that journeyed here as seeds, safely tucked away in the apron and trouser pockets of early colonists. Nettles, chickweed, dandelions and other wild greens emerging along with the grass and garden were once an eagerly anticipated spring treat. After a lengthy winter diet heavy in dried fruits, corn, beans and root crops, spring greens loaded with vitamins and minerals provided a refreshing lift to their to sluggish spirits and digestive systems.

Foraging for spring greens provides a firsthand opportunity to see the reawakening of nature after its long winter dormancy while getting fresh air and exercise the first warm days of spring. The plants described here are common weeds often found growing in vegetable gardens. *A Field Guide To Medicinal Plants, Eastern and Central North America*, by Steven Foster and James Duke, is an excellent field guide to correctly identify plants.

Nettle *(Urtica dioica)*

The nettle is a naturalized Eurasian perennial that can be found throughout most of the United States. Since ancient times it has been valued for its multitude of herbal uses. The Europeans and American Indians used the fibers from nettle stems to make sailcloth, cordage and fishing nets. The nettle has always been considered an excellent spring tonic and useful pot herb. The tender tops of new leaves do not sting as much and are highly nutritious, containing high levels of protein and iron. Historically it was used to treat rheumatism, dental problems and is still used today in hair products.

If you look closely at the plant's square stems and dark green leaves,

Nettle

they appear to possess a slight covering of short harmless hairs with a little bulge at the base. They are actually tiny tubes filled with formic acid. Anything brushing against the hair-like tubes will cause them to break into a jagged pointed stingers. "*He that handles a nettle tenderly, is soonest stung*" is an old adage that compares the nettle's unpleasant sting with those personality types, who in spite of being treated with sincere kindness, often fail to return the same courtesy to others.

Stinging nettles play an important role in both urban and rural wildlife. They provide food and serve as host plants for close to forty species of insects. Aphids winter over in nettle patches because they like to feed on the tender leaves in early spring. Ladybugs, a beneficial insect, depend on the aphids as an important early spring food source. Nettles are a favorite caterpillar host plant of the Red Admiral and Painted Lady butterflies for laying their eggs. Birds and small animals like to eat the dried nettle seeds.

Old time gardeners made an organic liquid fertilizer from nettles. They would fill a bucket with the tops of nettle plants and cover them with water. After steeping the plants for a few weeks, the plant material was strained out and the nettle-infused solution was used to fertilize choice plants.

Natural cordage or fibers made from nettle stems can be used to make friendship bracelets. For complete instructions and other foraging tips go to Natures Secret Larder, http://www.naturessecretlarder.co.uk/bushcraft-tutorials/nettle-cordage-tutorial.

Red Clover *(Trifolium pratense)*

Red Clover was introduced in this country from Europe, planted as a grazing and fodder crop to feed the colonists' horses, cattle and other domestic animals. It has become naturalized throughout most of temperate North America and can be found growing along the edges of forests, roadsides and in meadows and vacant lots across the United States. Red Clover is a biennial or short-lived perennial plant that grows in clumps, made up of several soft, hairy stems topped with plump reddish to rose-hued flower heads. A relative of the pea plant, it is among the largest of the numerous species of clover, reaching heights of two to three feet.

American Indians ate the leaves in early spring as a vegetable. The flowers were used to make wine, flavor cheese and brewed together with the leaves to make a healthful herbal tea. It has long been reputed for its medicinal value, and the Indians used the flowers to make an infusion to soothe sore eyes and as a salve for burns.

Red Clover has been associated with magical powers since ancient days. As night comes, the leaves on the side of the clover fold together and the center leaf protectively bends over them in a prayerful attitude. Travelers would carry a sprig of clover as insurance against witchcraft and evil spirits on their journeys. Legend states it is a symbol of domestic virtue and fertility, and clover leaves were used by Saint Patrick to represent the Holy Trinity.

Red Clover thrives in a variety of soils and is planted by farmers as a soil-enriching crop. Like other leguminous plants, red clover has nitrogen-fixing nodules on the roots. If you dig up a healthy clover plant and examine its roots, you will see the white nodules clustering among the plant's root hairs, evidence of the symbiotic relationship between the nitrogen-fixing bacteria and their gracious hosts, the clover roots. Red Clover was originally pollinated by bumblebees and was at one time called the bumblebee flower. Today it is more often pollinated by honey bees. The dried flowers make a lovely addition to potpourris and fresh clover blossoms, and with their stems attached can be fashioned into fra-

CLOVERS

The clovers have
no time to play;
They feed the cows
and make the hay.
And trim the lawns,
and help the bees,
Until the sun sinks
through the trees.
And then they lay
aside their cares.
And fold their hands
to say their prayers,
And drop their tired
little heads.
And go to sleep
in clover beds.
– HELENA L. JELLIFFE

grant ephemeral necklaces.

Purslane *(Portulaca oleracea)*

Purslane

Purslane, a native of India and Africa, was believed to be introduced into Europe as a salad plant during the fifteenth century. Early settlers brought purslane seed to America and planted it in their herb gardens. It has long been valued as an iron-rich spring green, a tasty vegetable and potherb by cultures the world over. Mahatma Gandhi was fond of purslane and encouraged his fellow Indians to cultivate it as a vegetable in the 1940s. An annual robust plant with a careless sprawling nature, succulent, round, fleshy stems and leaves, it flourishes in almost any soil. The nutritious crunchy, lemony-flavored leaves can be used raw or cooked and the mature stems were especially savored by past generations, pickled and spiced for winter use. Small yellow flowers blooming along the stems produce tiny brown pods filled with edible seeds that can be used much like poppy seeds. Purslane is the wild relative of the ornamental flower *Portulaca* sold in nurseries. This common weed readily springs up in gardens and waysides and makes a fine addition to early spring salads and coleslaw recipes. Purslane also makes a great groundcover in vegetable and flower gardens, helping the soil retain moisture in the heat of the summer, and therefore reducing the amount of watering that has to be done.

The Dandy Dandelion *(Taraxacum officinale)*

"Beauty is in the eye of the beholder." Some consider the dear dandelion a pushy, pesky weed to be banished upon first sight from neighborhood lawns and gardens, yet to others it is a bright harbinger of spring.

The name dandelion is from the French *"dent de lion"* meaning *"teeth of the lion,"* referring to the plant's jagged leaf edges. The Chinese called

Dandelion

them *"yellow flower earth nails,"* and years ago they were called *"shepherd's clocks"* because the flower opens at the beginning of the day and closes at dusk. Since dandelion flowers close on cloudy, rainy days, swine's snout is another of the plant's curious folk names.

The plant's proper name is *Taraxacum officinale* and is one of the eight species found in North America and the twenty-five growing worldwide. Most of our common spring flowers are short-lived ephemerals, but the delightful dandelion has one of the longest flowering seasons of any plant and can bloom all summer long. The plant is believed to be native to Asia Minor and arrived in the United States by way of Europe. Today it grows in temperate zones all over the world.

Dandelions have long been used for food, drink, medicine and by young children for lazy summer entertainment. They are still used around the world as a healthful spring salad green. Different parts of the plant can be harvested and eaten throughout the plant's life cycle. Before the buds appear in early spring, the young leaves can be harvested for use as salad greens and added to soups, eggs and baked goods. The spring root is tender and sweet and tastes like parsnips. Fried dandelion blossoms make a tasty spring snack. The roots can be harvested in the late summer or fall and used to make a strong, coffee-like drink.

American Indians simmered dandelion leaves with maple sap vinegar and mixed them with venison. The juice of freshly gathered flower stalks and leaves was used for removing warts.

Children have played with the gay, mop-headed dandelion for centuries, inventing games and creating crowns, chains and curls from dandelion stems and flowers. American Indian children used long dandelion stems to make whistles. At one time, every child knew how to tell time using a dandelion clock. You blow until the seed is all blown away; then you count each of the puffs, an hour to a puff.

CHILDREN'S FLOWER

Dear dandelion
you sunshiny thing,
How many toys
for young
folks you bring.

Watchchains for Nanny,
and trumpets for Ned
Funny green curls for
baby's bald head.

Next your white
seeds fly' which way
the winds blow,

Friend of the barefoot
boy, gold of the poor,
You're a wee play house
at every child's door.

– Author Unknown

GOOD FRIENDS — OUR NATIVE TREES

Flowering Dogwood *(Cornus florida)*

Hark, Hark,
The dogwoods bark.
The pussy willow meow,
The dandelion roars,
The hawkweed soars,
The cowslip moos
like a cow,
The Jack-in-The-Pulpit
tries to pray,
The Quaker ladies
all get gay,
The bluebells
start to ring!
The windflower blows,
The chickweed crows.
Its spring, spring, spring!

— EUELL GIBBONS

Flowering Dogwood is a beautiful understory tree native to Eastern North America. The blooming dogwood viewed at a distance in the spring woodlands, resembles a huge bouquet of white flowers, lovingly arranged and awaiting the return of spring. The dogwood tree has gone by other names throughout history — Indian Arrowood, Cornelian Tree, boxwood and dog tree. The family name of this group of trees, *Cornus*, is derived from cornu, meaning horn, and alludes to its extremely hard wood. Dogwood was used in times past to make tool handles, shuttle looms, jewelry boxes, spools and golf club heads. The common name dogwood has other origins. In previous times the name of an animal attached to a plant by early man was used to describe a plant part similar to that of the animal or perhaps was a favorite food of the animal. In olden days the astringent bark of the dogwood tree was reputed to be used as a cure for mange in dogs. American Indians simmered the bark in water and used the extract to relieve sore muscles. They also made a tea from the bark to break fevers. Flowering Dogwood was a favorite tree of George Washington and Thomas Jefferson. Both gentlemen mentioned its orna-

Flowering Dogwood

mental value in the home landscape in their personal papers.

One of the most popularly cultivated trees in Eastern North America, the dogwood is an attractive tree with a low branching habit and dense foliage. The true flowers are tiny, greenish-yellow clusters encircled by four white, showy bracts that look like large petals. When the Indians saw the dogwoods in bloom they knew it was safe to plant corn. The flowers are followed by clusters of shining red berries that provide a perfect accent to its scarlet autumn foliage. The tree is easily recognized in the winter by its distinctive purple-tinged twigs, interestingly textured gray bark patterned with small rectangular blocks and prominent square, gray flower buds tipping its branches. Flowering Dogwood grows in a variety of habitats and makes an outstanding four season accent tree in any size yard. It easily lends itself to up close observation and is a good tree for use when investigating plant life cycles and backyard wildlife. The twigs, bark, flowers and seeds are utilized as food by a variety of wildlife. Over 70 species of birds are known to feed on the fruit of various dogwoods. The little Spring Azure butterfly likes to lay her eggs on the leaves of the Flowering Dogwood.

Dogwood Blossom

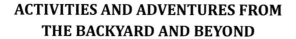

ACTIVITIES AND ADVENTURES FROM
THE BACKYARD AND BEYOND

It is important to use discernment and follow safe guidelines when foraging for wild greens.

• Before consuming any new plant gathered from the wild, make absolutely sure that you have identified it correctly and always get a second opinion. Beginners should only gather plants positively identified by a trained forager.

• Plants must be harvested a least 100 feet from the nearest roadside or highway to avoid contamination from car exhaust and other pollutants. City lots, urban fields and lawns may have been sprayed with herbicides, pesticides or fertilizers that can contaminate the plants growing there. Do not eat any plants unless you are certain they are free of chemicals.

• All wild-harvested greens should be soaked for 20 minutes in a gallon of cold water to which two or three tablespoons of vinegar or salt has been added. This soaking will remove the dirt, insects or wild things clinging to the greens. After soaking for 20 minutes they should be gently washed and rinsed with fresh water.

The following recipes are used with permission from Peter A. Gail, author of *The Dandelion Celebration* and *Volunteer Vegetable Sampler, Recipes for Backyard Weeds*. Both books have excellent, safe directions for foraging for wild greens in your backyard and beyond. *Volunteer Vegetable Sampler* contains a section on how to teach foraging to children, one step at a time.

For more recipes using plants, *Native Harvest, Recipes and Botanicals of the American Indians*, by Barrie Kavasch, is a treasure trove of information.

BAKED NETTLES AND POTATOES

5 medium potatoes
Salt and pepper
4 cups nettle tops
1 large onion, sliced
1 teaspoon nutmeg
Ginger
1 cup milk
2 ounces butter or margarine
Bacon

Generously butter a casserole dish. Slice potatoes as for scalloped potatoes. Season potatoes with salt and pepper. Add a layer of nettles and onion, season with nutmeg and ginger, ending with potatoes. Add about a cup of milk to come halfway

---

HOW TO MAKE DANDELION CHAINS

Pop the flower head from the dandelion stem. Bend the stem and insert the smaller end into the larger end to make a circle. Repeat, putting the next stem through the circle of the proceeding one. Keep adding stems until your chain reaches a nice length.

up the dish. Dot with plenty of butter or margarine and bake in a moderate oven (350°F) for one hour until done. Add lightly fried chopped bacon if you wish.

Bacon-Coated Purslane

6 slices of bacon
1/2 cup water
1/2 cup vinegar
4 teaspoons brown sugar
1 pound young tender purslane tips
Salt and pepper to taste
2 eggs, hard-boiled, finely chopped

Cut the bacon into small shreds and fry them until crisp. Add the water, vinegar and brown sugar. Simmer on low heat for 10 minutes. Stir constantly to prevent the sugar from burning. Clean and wash the young tender purslane. Add them to the mixture and stir gently so the tips are well coated. Season, garnish with the finely chopped eggs, and serve hot.

Clover Soup

2 cups clover blossoms and leaves, fresh or dried
2 small wild onions, chopped
4 tablespoons sunflower seed butter
1 quart water
3 medium potatoes, quartered,
Chopped fresh dillweed, or dried to taste
Salt and pepper

Saute the clover blossoms, leaves and the chopped onion in the sunflower seed butter. Add the water, potatoes and seasonings. Simmer covered for 20 minutes. Serve hot.

Dandelion Pizza

Use your favorite dough recipe
4 cups finely chopped dandelion leaves
1 clove garlic, minced
2 tablespoons Parmesan cheese
Juice of 1/2 lemon

Shredded mozzarella cheese

Olive oil

Prepare pizza dough using whatever recipe you prefer, and let rise. When ready to bake, place on a baking sheet in 350°F oven and bake until golden, but not brown, about 12 minutes. Remove from oven and brush on olive oil with pastry brush. Let cool slightly. Steam dandelions and drain well. Lightly sauté the garlic in olive oil and add greens. Mix Parmesan cheese and lemon juice. Spread greens on baked crust and top with shredded mozzarella cheese, Parmesan cheese and lemon mixture. Return to oven for about 10 minutes or until cheese melts.

DANDELION COOKIES

1/2 cup vegetable oil

1/2 cup honey

2 eggs

1 teaspoon vanilla extract

1 cup unbleached flour

1 cup dry oatmeal

1/2 cup dandelion flowers

Blend honey and oil and beat in 2 eggs and vanilla. Stir in flour, oatmeal, eggs and dandelion flowers. Drop the batter by teaspoonfuls onto a lightly-oiled cookie sheet. Bake in preheated 375°F oven for 10 to 15 minutes.

Dyeing Easter Eggs Naturally

The egg has always been used to symbolize new life. The custom of dyeing eggs for spring holiday celebrations dates back to antiquity. Try coloring your Easter eggs using natural ingredients.

DYEING INSTRUCTIONS

Suggested plant materials to use for coloring eggs:

Brown – onion skins

Red – pomegranate juice

Blue – red cabbage leaves or blueberries

Yellow – yellow mustard or turmeric

Purple – concord grapes

Pink – fresh shredded beets or raspberries

Silver Green – purple basil

Do one color at a time. Place 6 to 12 eggs in a large pan. Cover the eggs with three cups fresh, dried or frozen plant material. Cover the dye material and eggs with water and add two tablespoons of vinegar. Bring to a boil and then gently simmer for ten minutes. Remove from heat and let eggs set in the dye water until cooled. Store in the refrigerator.

Recommended Reading

Gail, Peter A. *The Dandelion Celebration*. Cleveland, Ohio: Goosefoot Acres Press, 1991.

Gail, Peter A. *Volunteer Vegetable Sampler: Recipes for Backyard Weeds*. Cleveland Heights, Ohio: Goosefoot Acres Press, 1991.

Foster, Steven and James. Duke A. *A Field Guide To Medicinal Plants, Eastern and Central North America*. Boston, Houghton Mifflin Company, 1990.

Kavasch, Barrie. *Recipes and Botanicals of the American Indian*. New York: Dover Press, 2005.

THYMELY TIPS AND SAGE ADVICE

• Dogwood winter describes a spell of cold weather and heavy frost that frequently follows a lovely spell of warm weather in the midst of spring, about the time the dogwood trees are in bloom. This is usually between mid-April and early May.

• *Nature Study in Your Own Yard* is a charming collection of nature-study leaflets written by Liberty Hyde Bailey and Anna Botsford Comstock, published through the College of Agriculture at Cornell University

TREES

I think that I shall
never see
a poem lovely
as a tree.
A tree whose mouth
is pressed against
the earth's sweet
flowing breast:
A tree that looks at
God all day, and lifts her
leafy arms to pray;
A tree that may in
Summer; wear a nest
of robins in her hair;
Upon whose bosom
snow has lain;
who intimately
lives with rain.
Poems are made by
fools like me, but only
God can make a tree.

– JOYCE KILMER

When we try to pick
out anything by itself,
we find it hitched to
everything else
in the Universe.

– JOHN MUIR

between 1864-1904. The leaflets were created to aid teachers and students studying plants. They are delightful and informative to read. *A Study of a Tree Leaflet #47*, by Anna Botsford Comstock, would be great to use for Arbor Day. They are available online at The Liberty Hyde Bailey Museum, http://lhbm.south-haven.com.

• The first Arbor Day took place in Nebraska on April 10, 1872. After Julius Sterling Morton moved to Nebraska from Michigan and noticed the lack of trees, he decided to set an example by planting orchards, shade trees and windbreaks on his property and urged his friends and neighbors to do the same. When serving on Nebraska's Board of Agriculture, Morton organized a state-wide day dedicated to tree planting and education, demonstrating the positive effect of trees in our daily lives. In 1970 it was designated a national holiday and is celebrated across the United States, www.arborday.org. It's a good time to take stock of the location and health of the trees in your yard and neighborhood. The National Tree Benefit Calculator, or i-Tree, http://www.itreetools.org, is a program developed by the U.S. Forest Service designed to help cities increase citizen awareness of the ongoing benefits trees provide for the community. If you visit the Web site, it will show you how to calculate the value of trees in your yard and neighborhood.

• On April 22,1970, the first Earth Day was celebrated in the United States. It was a day intended to inspire awareness and appreciation for the Earth's natural resources. Founded by U.S. Senator Gaylord Nelson as an environmental education day, it is now celebrated in more than 175 countries every year.

• John Muir, Scottish-born American naturalist and founder of the Sierra Club, was born on April 21,1838. The Sierra Club has resources and information for communities, schools and individuals who would like to celebrate John Muir's birthday, www.sierraclub.org/john_muir_exhibit.

MAY

Merry, rollicking, frolicking May

Into the woods came skipping one day;

She teased the brook till he laughed outright,

And gurgled and scolded with all his might;

She chirped to the birds and bade them sing

A chorus of welcome to Lady Spring;

And the bees and butterflies she set

To waking the flowers that were sleeping yet,

She shook the trees till the buds looked out

To see what the trouble was all about;

And nothing in nature escaped that day

The touch of the life-giving, bright young May.

— GEORGE MacDONALD

Flower Moon — Planting Moon — Milk Moon

American Indians were experts at reading the landscape and their knowledge of nature's calendar helped ensure their survival. They appreciated and respected the interdependence of all living things. Their lives revolved around seasonal cycles and patterns of movement as they moved from one food source to another, continually modifying their behavior in response to the life cycles of local plants and animals.

The Flower Moon signaled it was time to begin scheduling the summer's tasks, activities and ceremonies. By closely observing blossoming plants and the wildlife interacting with them, the Indians were able to predict the best time to plant their gardens, harvest wild berries and collect natural materials for making tools and household articles.

Previous experience had taught them the best time to gather basket grass and cedar roots was when the wild rose bloomed. White Oak leaves the size of a mouse's ear indicated the time to plant corn was at hand. By following seasonal rounds or cycles, they were able to make efficient use of available food and resources within their local regions in a timely manner. The order in which the wild plants bloomed throughout spring and summer helped them organize their daily tasks. Flowers vary in their bloom time in different places, but they all do bloom about the same time in any given place each year.

Phenology is an interdisciplinary branch of ecology that uncovers the balance and mutual exchange taking place daily in the natural world. Children can understand its core concepts and scientists never cease to be amazed by the insights it generates. Native plants occurring naturally to a geographical region are more likely to possess interesting adaptations to local climate and ecological conditions than plants introduced and artificially maintained. It is best to focus on natives when observing plants to learn how they respond to seasonal changes.

OJIBWA
SPRING SONG

As my eyes
look over the prairie
I feel the summer
in the spring.

From a translation in
Frances Densmore,
Chippewa Music II.
Bae Bulletin 53, 1913

Wild Rose

Natural history observations found in ancient records and literature dating back to biblical times reveal a common level of understanding about phenology among early peoples. *"Learn a lesson from the fig tree. Once the sap of its branches runs high and it begins to sprout leaves, you know summer is near."* Gospel of Mark 13:28

The Planting Moon marked the beginning of the gardening season. Each family was responsible for their own plot of land and the crops they produced belonged to them. Planting and tending their gardens allowed them to honor and show respect for laws of nature while actively participating in the circle of the seasons. In the spring they cleaned and burned off the dead debris in their gardens and then prepared the soil and planted the seeds, saved from previous year. Throughout the summer while keeping their gardens, they would sing to the growing plants and give thanks for the blooming flowers.

Planting corn, beans and squash together was a traditional practice of American Indians; they called them the three sisters. The corn was planted first, and after it started to grow they planted squash and bean seeds around the base of the corn stalks. The squash shaded the roots of the corn and beans would grow up the corn stalks. Corn was an important staple in American Indian cultures. It was used for food, baskets, moccasins and dolls were made from the cornhusks. The corn silk was used medicinally and they used the corn kernels as beads to create beautiful jewelry. Corncobs were utilized for fuel and darts.

According to American Indian folklore, man corn was looking for a wife and considered squash at first, but she wandered about too much. When bean firmly clasped the corn stalk assuring her fidelity, corn chose her as his bride. It was a wise choice because legumes have nitrogen-fixing bacteria that help grow strong and healthy corn plants. Farmers still practice a form of this type of companion planting when they rotate corn

Though I do not believe that a plant will spring up where no seed has been, I have great faith in a seed. Convince me that you have a seed there, and I am prepared to expect wonders.

– HENRY DAVID THOREAU

crops with soybeans every other year. There are about 45,000 items in the average American supermarket, and more than a quarter of them contain corn. It truly is an amazing plant!

Some of the Eastern Woodland tribes referred to this period of the year as the time of the Milk Moon, when many types of domestic livestock, as well as wild animals, give birth and nurse their babies.

Wildflowers

Wildflowers are blooming plants that can survive and flourish without any help from humans. Over 10,000 types of wildflowers exist in North America. They can be found independently thriving in a variety of habitats ranging from wet swamps, roadside ditches, woods, meadows and water routes across America. Many of them are natives, beautiful fragments of a region's natural history, present here long before European settlement. Others are green immigrants that were imported here deliberately or accidently from Africa, Asia or Europe. Some of them come and go each year before we even realize they have made their seasonal appearance. People often value them for their beauty first, and their role in the environment second. Wildflowers are important members of the natural communities where they have evolved over thousands of years. Early man valued many wildflowers for their flowers, seeds, fruits and roots for use as food and medicine.

All plants have an internal biological clock that helps them gear up for the seasons and climate conditions. It helps them measure the passage of time, respond to the environment and make efficient use of their energy. We do not fully understand how plants know what time of day or season of the year it is, but we do know that plants respond to changes in light and temperature. And some of us know it is great fun to observe the plants and flowers associated with the different months, seasons and days of the year.

Spring ephemerals are short-lived wildflowers that must begin and end their growth cycle during a limited time in early spring, when the ground is warm enough for growth and the trees have not leafed out enough to block the vital rays of the sun. Herbaceous non-woody plants, they quickly disappear beneath the natural debris covering the ground after blooming each spring. They use their leaves to make food through the process of photosynthesis, which they promptly store away in thickening roots, bulbs or underground stems hidden beneath the soil surface.

This food allows spring wildflowers like Spring Beauty, bloodroot, trout lily and Jack-in-the-Pulpit, to jump into action during the first warm days of spring. That's why picking early blooming wildflowers ultimately kills them. They need their leaves to manufacture their food, making it possible for them to survive through winter and bloom again in spring.

Do not be fooled by their apparent haste to bloom in early spring. There is more here than meets the eye. Trilliums, one of our most well-known perennial wildflowers, can take up to six years before it produces its first blooms. Trilliums grow in rich, moist woods and thickets throughout North America. The flower's name describes the plant's parts; it has three leaves, flowers with three petals and three sepals. Trilliums bloom about the same time the robin returns and the earthworms begin moving about beneath the soils surface, hence the common name Wake Robin. The Red Trillium, the Showy Trillium and the Nodding Trillium are the most common of the thirty species of trilliums.

The Red Trillium is the first to bloom in early spring, emitting a rank odor which attracts flies and beetles that help pollinate its stinky flowers. The butterflies and bees prefer the Showy Trillium which comes into bloom a bit later in spring. In midsummer, when the plant's fleshy red berries are ripe, ants come along and eat the berries' sweet covering and

The world may find
spring by following her,
Here was she
wont to go!
and here! and here!
Just where these daisies,
pinks and violets grow,
The world may find
spring by following her.
And where she went
the flowers took the
thickest root.
– Ben Johnson

Bloodroot

then help plant the seeds. A trillium berry is star-shaped with three ribs and is very interesting to observe with a hand lens.

American Indians used the plant medicinally and concocted love potions from the boiled root. The Indians have a story about a beautiful young maiden who was set on winning the heart of a handsome young warrior and becoming his wife. She boiled a trillium root to add to his food so he would have eyes for no one but her. But when she was trying to creep undetected to his wigwam, she tripped and the root fell into the food pot of a cranky old man, who quickly gobbled it up and followed the unlucky girl around for months, begging for her hand in marriage.

Trilliums are a great choice for the beginner wildflower gardener and they are readily available from established wildflower nurseries and native plant suppliers. Every home landscape can benefit from the addition of a few wildflowers tucked in out of the way places or planted in established garden beds. Most of our common wildflowers can be successfully cultivated and once they become established they need little maintenance. Every plant has an interesting past, its own legends and uses by man and wildlife. Learning the details of a plant's personal life make them seem like close friends.

A Guide to Enjoying Wildflowers, by Donald and Lillian Stokes, is a great guide for following wildflowers through the seasons as well as histories of their use.

When out and about looking for wildflowers, find an old log or large rock to sit on. Listen and look for signs of birds and animals inhabiting the area. You can clearly see the nesting cavities in trees, entrances to animal homes in the ground and evidence of wildlife around brush piles and fallen trees. Take along a notebook and make a list of things you notice and save it to compare with your next visit later in the month. Sketch

Trillium

Showy Trillium

Nesting cavity in old log

the wildflowers and trees. If you were a bird or animal looking for a home where would you choose to live in this wildlife community?

The Handbook of Nature Study, by Anna Botsford Comstock, is close to a century old and is still one of the best nature studies available today. Every family should have a copy of this classic on their library shelf. An exceptional collection of nature lessons, it is a must-have resource for anyone working with children's gardening or nature study programs. *The Outdoor Hour/Nature Study Close to Home* is an award-winning, interactive Web site based on the book with ideas on how to spend time outdoors for an hour each day. Check it out at www.handbookofnature study.blogspot.com.

Viola

Violets and pansies belong to the same family Violaceae and genus *Viola*. Violets and pansies are frequently available for planting in spring and can be used in a variety of spring recipes, games, science projects and homemade crafts.

There are about seventy-five species of violets in North America, a large number of them native. Some have white or yellow flowers, but the Blue Violet is the most familiar. Violets are divided into two distinct groups: plants with straight stems and leaves and flowers on them, such as the Johnny-Jump-Up *(V. tricolor)* or those like the common Blue Violet *(V. papilionacea)* whose leaves and flowers grow from underground rhizomes or stolons. American Indians used these rhizomes to make an infusion to soak their corn seed before planting to repel insects.

The Blue Violet returns each spring about the same time as the redbud tree begins to bloom, and they are frequently found growing in the same vicinity. Another similarity shared by the redbud and violet is their heart-shaped leaves. But, unlike the redbud, the leaves of the violet emerge be-

Plant a flower,
Grow a friend.

fore the flowers. The fresh green leaves unroll from the base of the plant as they mature. Violet leaves and flowers have been used for food and medicine since ancient times.

The plant's young leaves have long been hailed as a healthy spring green because of their high levels of vitamins C and A. The taste of violet leaves has been compared to the flavor of spinach. At one time the flowers were used to make candy, jam and syrups. Chemists once used the juice from mashed violet flowers to determine the acid or alkaline nature of a substance. When violet juice is touched by an acid the juice turns red; if touched by an alkaline substance it turns green.

Humans aren't the only ones who crave the fresh green leaves. Cutworms, the larva of noctuid moths, cut the leaves off and eat them; slugs chew holes shaped like irregular circles in the leaves at night and hide beneath them during the day. Violets are the caterpillar host plant of the Great Spangled Fritillary and the Lesser Fritillary Butterfly.

Violet flowers possess a fun adaptation that bees and butterflies find quite accommodating. The flower has five sepals and five easily seen petals. If you examine the base of the lowest petal you will notice it projects farther back than the base of the flower into a spur-like well that holds the nectar. The elongated bottom petal acts as a landing pad for insects stopping by for nectar. In case cold temperatures or unpredictable weather prevent insects from pollinating the pretty blue flowers in spring, the plant produces tiny inconspicuous self-pollinating flowers hidden beneath leaves close to the ground to ensure seed production for the next year. Ants prize the sweet outer shell of the seeds as treats. They carry them back to their nests, eat the covering and help plant them when they discard the seed in soil beneath the ground.

American Indian children played games with violets. They would divide into two groups, one group taking the name of their tribe and the

He who bends a knee
where violets grow,
a hundred secret things
shall know.
– Anonymous

Books in running brooks,
Sermons in stones and
good in everything.
As it fell upon a day,
In the merry
month of May.
– Richard Barnfield

other would take the name of a different tribe living close by. Each team collected an equal number of long-stemmed violets and sat down facing each other. Each player would take a violet in hand and hook the head of the violet flower under the head of his opponent's flower. The projecting spur under the curved stem at the base of the violet worked like a hook clinching the two flower heads together. They would pull until the stronger conquered the weaker by popping off the head of his flower. Whichever tribe accumulated the most flower heads won the game and would taunt the losing team by calling them weak fighters.

Pansies are descendants of the small Johnny-Jump-Up, and both are popular garden favorites, introduced to America from Europe. People have been planting them in their spring gardens for hundreds of years. The violet is sometimes called the sister of the pansy. All three flowers have a wealth of lovely poems, legends and variety of names associated with their origins. The violet symbolizes constancy and modesty, and its history dates back to ancient Greece and Rome. The pansy name is a corruption of the French word *pensees*, meaning thoughts, and pansies have long been associated with thoughts of love.

"Anyone in close sympathy with flower and tree and shrub and has a general acquaintance with Nature's moods, could tell the time of year without any reference to a calendar."

– Gertrude Jekyll, *Home and Garden*, 1900

Old time gardeners relied on the bloom sequence of common plants growing in their local regions to let them know it was safe to sow seeds and set plants into the garden. The chart below is based on a copy of a list landscape architect Nelva M. Weber produced for the April 1947 issue of *The Home Garden Magazine*.

PANSY PREDICTIONS

The pansy has long been used to tell people's fortunes. Randomly pick a petal off a pansy and look at its markings:

– Four lines are a sign of hope.

– Five lines from a center branch are hope founded in fear.

– Thick lines bent to the right mean prosperity.

– Thick lines bent to the left mean trouble ahead.

– Seven streaks mean constancy in love (and if the center streak were the longest, Sunday would be the wedding day).

– Eight means fickleness.

– Nine means a changing heart.

– Eleven is disappointment in love and an early grave.

– Traditional Lore

Homemade Time Table for Seed Sowing

When you see the flowers of these plants in bloom:

Spice Bush *(Lindera benzoin)*

Border Forsythia *(Forsythia intermedia)*

Siberian Squill *(Scilla sibirica)*

These seeds may be safely planted:

Emerging Seed

HERBS & VEGETABLES	ANNUAL FLOWERS
Beets	Calliopsis
Cabbage	Candytuft
Chard	Mignonette
Chervil	Rudbeckia
Kohlrabi	Snapdragon
Lettuce	
Onion (sets, seed)	
Parsnips	
Parsley	
Peas	
Spinach	
Turnips	
Leeks	

When you see the flowers of these plants in bloom:

Juneberry *(Amelanchier canadensis)*

Moss Phlox *(Phlox subulata)*

Bleeding Heart *(Dicentra spectabilis)*

These seeds may be safely planted:

And with your child you can ponder the mystery of a growing seed, even if be only One planted in a pot of earth in the kitchen window.
— RACHEL CARSON

HERBS & VEGETABLES	ANNUAL FLOWERS	
Broccoli (seeds or plants)	Feverfew	Marigold
Cauliflower (seeds or plants)	Annual Phlox	Zinnia
Carrots	Nicotiana	Pansy
Celery	Calendula	Lobelia
Fennel	Petunia	Pinks
Potatoes	Cleome	Scabiosa
Dill	Four O'clock	Stock
Borage	Gaillardia	

When you see the flowers of these plants bloom:

Redbud *(Cercis canadensis)*

Flowering Dogwood *(Cornus florida)*

Lilac *(Syringa vulgaris)*

These seeds may be safely planted:

HERBS & VEGETABLES	ANNUAL FLOWERS
Corn	French Marigold
Cucumbers	Morning Glory
Muskmelons	Nasturtium
Pumpkin	Salpiglossis
Snap Beans	Statice
Squash	Sunflower
Zucchini	Bachelor Button
Sweet Marjoram	
Summer Savory	

When you see the flowers of these plants in bloom:

Perennial Flax *(Linum perenne)*

Vanhoutte Spiraea *(Spiraea* x *vanhouttei)*

Japanese Snowball *(Viburnum tomentosum var. plicatum)*

The seeds of these may be safely planted:

HERBS & VEGETABLES	ANNUAL FLOWERS
Bush Lima Beans	Ageratum
Scarlet Runner Bean	Coreopsis
Soybeans	
Hyacinth Bean	

When you see the flowers of these plants in bloom:

Common Peony *(Paeonia officinalis)*

Rugosa Rose *(Rosa rugosa)*

Highbush Cranberry *(Viburnum opulus)*

These seeds and plants may be safely planted:

VEGETABLES

Eggplant (plants)

Okra

Pepper (plants)

Pole Lima Beans

Sweet Potatoes (plants)

Tomato (plants)

Watermelon

Basil

ANNUAL FLOWERS

Dahlia (plants)

Portulaca

Salvia

IN THE HEART
OF A SEED

In the heart of a seed,
Buried deep, so deep,
A dear little plant
Lay fast asleep.

"Wake" said the
sunshine,
"And creep to the light,"
"Wake" said the voice of
the rain drops bright.
The little plant heard,
And it rose to see

What the beautiful
Outside world might be.

— Kate L. Brown

Planting Seeds

Nothing beats the magic of watching a tiny seed push up through the soil and the joyful satisfaction of harvesting tasty homegrown vegetables, fragrant herbs and colorful flowers. Seeds come in all shapes, colors, textures and sizes, each one representing both the beginning and ending of a complete life cycle. The choice of plants and varieties available in seed far outshine the plant selections found at most garden centers. Seed catalogs provide useful descriptions of plants and cultural information. Seed packets give detailed instructions for successfully sowing seeds. There are a few essential points to ensure a successful experience when sowing seeds in the garden.

Be weather wise. There are cool season and warm season plants. Some plant species are hardy enough to go into the ground as soon as the frost is past and the ground can be worked. Others must wait until all danger of a late frost is gone.

Be soil wise. Follow this simple time-honored test for determining whether or not soil is ready to be worked in the spring and safe for planting seeds. Grab a handful of soil and squeeze it into a compact ball or lump. If it remains a solid sticky lump, the ground is still too wet to plant. If the lump of soil easily crumbles apart, it's okay to begin planting

Start from the ground up; make sure your soil is properly prepared. Most annuals like a good loamy soil. This is a combination of clay, organic matter, silt and sand. Organic gardening is a system of gardening that utilizes nature's methods of enriching the soil, returning more to the soil than the plants use. Well-amended soil holds water longer and dries out more slowly. Adding compost helps preserve the humus and is healthier for the soil than using chemical fertilizers. Organic fertilizers have more nutrients and release them slowly, depending on the needs of the plant.

Do not sow seeds too thickly or deeply; seeds should be planted according to their ultimate size and growth habit. Check the seed packet for directions about seed depth and planting times. Mix and mingle the slowpokes with other quick-germinating plants so children will always have something to watch and tend. Tuck some seeds in a few out-of-the way places or crevices around your yard and neighborhood. Then watch them grow and compare how they respond to the varied growing conditions.

Be sure to thin out young seedlings, giving each one ample room to grow and reach their proper size. Transplant extras in other locations in the garden or share them with your neighbors. When transplanting young seedlings, help direct the rainwater to the plants root zone by creating a shallow indentation in the soil around the base of the plant.

It is very important to keep the soil evenly watered so the seeds don't dry out. Water on a regular schedule during dry weather.

Mist in May and heat in June make the Harvest right soon.
– TRADITIONAL LORE

GOOD FRIENDS — OUR NATIVE TREES

Redbud *(Cercis canadensis)*

The Eastern Redbud is a small understory tree found growing in the wild from the Central Atlantic coastline into the Midwest. The tree has an upright trunk that divides into sturdy branches topped with a flat spreading head if given an unhindered space to grow. The redbud belongs to the pea family Leguminosae and the genus *Cercis*, the name is

Redbud

attributed to the shape of the tree's fruit, which supposedly resembles a tool used by weavers in ancient Greece.

The redbud has long been valued as a season indicator because it blooms in early spring about the same time as the dogwood and Juneberry. Tiny clusters of rose-colored, pea-shaped blossoms emerge before the leaves, breaking out on the old wood, new twigs and along the trunk on some older trees. In times past, children enjoyed eating them in early spring. Susan Tyler Hitchcock in her book, *Gather Ye Wild Things*, recommends tossing redbud blossoms over a spring salad or stirring them into custard or rice pudding for a bit of cheery color and fresh lemon flavor. Because the early, bright pink flowers always appeared before the leaves the American Indians would say *"pink flowers form into leaves."* They would bring blossoming branches of redbud into their homes to drive out winter.

An interesting legend arose about the tree because of the unusual habit of the pink blossoms breaking out on old branches and along the trunk. When early settlers saw it glowing pink in the leafless woodlands in early spring they said it was blushing in shame because Judas hung himself on a redbud tree. The specific species of redbud that was originally labeled the Judas tree grows in the Mediterranean and Asia Minor countries.

The tree has decorative heart-shaped leaves that turn a bright clear yellow in the fall. It makes a lovely shade tree in small gardens and compact yards. The tree looks good in groupings in a lawn or on the edges of naturalized areas. They tolerate partial shade but bloom more profusely when provided a sunny location. Redbud trees can flourish in a variety of soil types, excluding wet and swampy soil. Many cultivars of this happy tree have been developed so gardeners have a nice selection to choose from to satisfy their needs. The seeds are enclosed in flat thin pods and are a valuable source of food for birds and deer. The Henry's Elfin But-

MAY DAY MORNING

Oh let's leave a basket
of flowers today,
For the little old lady
who lives down our way.
We'll heap it with
violets, white and blue,
With a Jack-in-the-Pulpit
and wind flowers, too.

We'll make it of paper
and line it with ferns,
Then hide – and we'll
watch her surprise,
When she turns and
opens her door and
looks out to see
Who in the world it
could possibly be!

– Virginia Scott Miner

terfly uses the redbud as a caterpillar host plant for laying her eggs.

ACTIVITIES AND ADVENTURES FROM THE BACKYARD AND BEYOND

The first day of May is observed in many countries with all sorts of outdoor festivities celebrating the arrival of spring. May Day celebrations and traditions can be traced back to the Roman festival of the Floralia honoring the goddess of springtime Flora. Some people believe the celebration dates to the Druids and their worship of trees.

The Romans celebrated the day with flower-decked parades. In England, May poles were erected on village greens on the first morning in May. The village youth gathered wildflowers and hawthorn blossoms to trim May poles. They would choose a king and queen to reign over the festivities. Villagers would dance and sing round the May pole holding the ends of colorful ribbons streaming from its top. The ribbons were wound around the May pole until it was covered in bright colors. In some countries it was considered an official day for courting. In Italy, boys serenaded their girlfriends with love songs. In Switzerland and Germany, young men would plant trees beneath the windows of their sweethearts. In France, they dedicated the day to the Virgin Mary. May Day has never been celebrated as an official holiday in the U.S. Our Puritan forefathers frowned on May Day festivities. It was banned in England in 1644 and was not resumed until 1660.

One delightful tradition that has survived is the making and giving of May Day baskets. Most often they are made of paper in the shape of a cornucopia with a handle and filled with the first flowers of spring. The baskets hung on the front door of a relative or neighbor as a token of love and friendship. Take some time to remember a neighbor or friend with a token of friendship, love, and remembrance — a May basket.

SIMPLE MAY BASKET

Materials needed:
- Tacky glue
- A 8-inch x 10-inch sheet of waxed freezer paper.
- A 8-inch x 10-inch sheet of wall paper.
- Decorative edging shears
- Paper punch
- Ribbon

May Basket

Glue the rough side of waxed freezer paper to the back side of the wall paper sheet and let dry. Trim the edges of the paper with the decorative edging shears. Roll the sheet of lined wall paper up into a cornucopia. Lap over and paste the paper's edge to the side of the basket. Use the paper punch to make a hole on each side of the basket to attach the ribbon for the handle. Fill with cultivated spring flowers and ferns and hang on a friend's door.

PLANT A TALL GRASS ZOO

Homeowners are being encouraged to leave patches of natural space on their properties for wildlife. If you don't want to leave an unmowed patch of lawn, consider planting a little garden of grasses in an out of the way corner of your property.

There are a wide range of perennial native grasses available in different heights, all with attractive flower heads. Some are cool season plants that bloom in early summer; others are warm season grasses that bloom in late summer and early fall. Many annual grasses have showy flower heads that can be used dried for winter arrangements. They are low maintenance, easy to grow and can provide both food and shelter for an amazing number of beneficial insects and animals — spiders, beetles, grasshoppers, various caterpillars and toads. The seed heads that form on grasses in late summer and early fall provide food for winter birds. Keep a record of the insect and wildlife activity taking place throughout the different seasons.

CRYSTALLIZED SPRING FLOWERS

Delicate edible spring violets, pansies and tiny flavorful leaves of newly emerging herbs such as mint, sweet cicely, lovage and strawberry make beautiful tasty decorations for cakes, cookies and ice cream.

Materials needed:
- Freshly picked violets, pansies and leaves of herbs
- Egg white and a few drops of water
- Paint brush, wax paper and a pair of long tweezers
- Superfine sugar in a salt shaker

Pick the flowers and herbal leaves in the morning after the dew has dried. Quickly rinse the leaves and flowers in cold water and lay them on a paper towel to drain. Whisk an egg white with a few drops of water. Using the tweezers to hold the leaves or flowers, dip them in the egg white and shake the sugar over the flower or leaf. Some people prefer to use a paint brush. Place each sugared flower on wax paper to dry for two or three days. Store in air-tight jars for up to one year.

Nourishing Nature Naturally

Now is the time to set out your birdbath. Be sure to locate it at least fifteen feet away from low growing shrubs that might provide the perfect cover for hungry cats. A clean water source is necessary for birds living and nesting in your gardens and yard.

Birdbaths should be no deeper than three inches at the deepest point so the bird can wade and splash around comfortably. The surface should be rough with an edge around the top so the birds have a place to perch. Be sure to keep it clean and free of algae. Change the water every few days.

Birds are actively engaged in nest building from May through August. Listen and watch for baby birds hatching in nests and birdhouses in your backyard and beyond. Be sure to contribute the bird species and your observations about their family activities to *NestWatch*, www.birdsource.org. The American Birding Association, www.americanbirding.org has a checklist of birds you can copy and use to record the different

A swarm of bees in May,
Is worth a load of hay;
A swarm of bees in June,
Is worth a silver spoon,
But a swarm
of bees in July,
Is not worth a fly.
– Traditional Lore

species coming and going throughout the year. They also have some great online materials designed especially for children. Check out the American Birding Association Young Birder's Home Page, http://www.aba.org/yb.

THYMELY TIPS AND SAGE ADVICE

• Rachel Carson, biologist, ecologist and writer, was born in Springdale, Pennsylvania on May 27,1907. She is considered the mother of the modern day environmental movement. Her book, *Silent Spring*, published in 1962, led to the banning of DDT, the creation of the Clean Water Act and the development of the Environmental Protection Agency (EPA). She passed away in 1964 after a long battle with breast cancer. She remains one of the most influential nature writers of all time. In 1950 she wrote an amazing magazine article called *Help Your Child to Wonder.* The piece was written to encourage parents to spend time nurturing their children's inborn sense of wonder about the natural world. The work was republished in book form and retitled *The Sense of Wonder.* Check out the following Web sites for more information and resources: www.rachelcarsonhomestead.org and www.rachelcarson.org.

• Late spring and early summer is the best time to prune spring flowering shrubs. These trees and shrubs usually bloom next spring on the stems produced during this growing season. Try to finish before the 4th of July. Save some of the twiggy branches to use as supports for low-growing weak stem plants in your garden.

• May is a good time to plant biennial and perennial seeds in the garden for flowers next year. A biennial plant takes two years to flower and set seed. Perennials have a lifespan over two years. A good rule of thumb to remember when planning perennial gardens: the width of the garden should be about twice the height of the tallest plant growing in it.

SPRING SOUNDS

Cracking ice
Dripping water
Peeping & Croaking
Pelting rain
Sloshing feet
Bubbling brooks
Singing birds
New life
Blowing wind

PATTERNS

Fragile flowers
Breaking buds
Seeds & Eggs
Underground roots
Green blades
Soft edges
Tiny leaves
Hearts

• Tuck some night-blooming plants in and around your patio or deck. Many of the plants that bloom at night have fragrant white flowers. Night bloomers attract night-flying moths that feed on their nectar and pollen.

• The Branch County Herb Questers, in Coldwater, Michigan, recommend planting these herbs from seed directly into the ground after all danger of frost is past: dill, coriander, parsley, chervil, salad burnet, summer savory, fennel, calendula and ambrosia. These herbs readily self-sow, so allow at least one plant to make seed. When the seed is ripe, shake it onto an area that will not be disturbed in early spring.

• Weed, fertilize and mulch your gardens. While weeding, keep an eye out for insect and critter damage. *"A weed that runs to seed, is a seven years' weed."*

Summer

Then followed that beautiful season… Summer…
Filled was the air with a dreamy and magical light;
and the landscape
Lay as if new created
in all the freshness of childhood.

– Henry Wadsworth Longfellow

HOW THE SUMMER CAME

Morning Glory was tired of the winter and longed for the spring to come. Sometimes it seemed as if Ka-bib-on-okka, the fierce old North Wind, would never go back to his home in the Land of Ice. With his cold breath he had frozen tight and hard the Big-Sea-Water, Gitche Gumee, and covered it deep with snow until you could not tell the Great Lake from the land.

Except for the beautiful green pines, all the world was white, a dazzling, silent world in which there was no musical murmur of waters and no song of birds.

"Will O-pee-chee, the robin, never come again?" sighed Morning Glory. "Suppose there was no summer anywhere and no Sha-won-dasee, the South Wind, to bring the violet and the dove. O, Iagoo, would it not be dreadful?"

"Be patient, Morning Glory," answered the old man. "Soon you will hear Wa-wa, the wild goose, flying high up, on his way to the North. I have lived many moons. Sometimes he seems long in coming, but he always comes. When you hear him call, then O-pee-chee, the robin, will not be far behind."

"I'll try to be patient," said Morning Glory, "but Ka-bib-on-okka, the North Wind, is so strong and fierce. I can't help wondering whether there ever was a time when his power was so great that he made his home here always. It makes me shiver to think of it!"

Iagoo rose from his place by the fire and drew to one side the curtain of buffalo hide that screened the doorway. He pointed to the sky—clear and sparkling with stars.

"Look!" he said. "There, in the North. See that little cluster of stars. Do you know the name we give it?"

"I know," said Eagle Feather. "It is O-jeeg An-nung, the Fisher stars. If you look right, you can see how they make the body of the Fisher. He is stretched out flat with an arrow through his tail. See, sister!"

"The Fisher," repeated Morning Glory. "You mean the furry little animal, something like a fox? Is Marten another name for it?"

"That's it," said Eagle Feather.

"Yes, I see," nodded Morning Glory, "but why is the Fisher spread out flat that way in the sky, with an arrow sticking through his tail?"

"I don't know just exactly why," admitted Eagle Feather. "I suppose some hunter was chasing him. Perhaps Iagoo can tell us."

Iagoo closed the curtain and went back to the fire.

"You thought there might have been a time when there was no summer on the earth," he said to Morning Glory, "and you were right. Until O-jeeg, the Fisher, found a way to bring the summer down from the sky, the earth was everywhere covered with snow and it was very cold. If O-jeeg had not been willing to give his life so that all the rest of us could be warm, Ka-bib-on-okka, the North Wind, would have ruled the world as he now rules the Land of Ice."

Then Morning Glory and Eagle Feather sat down on the soft rug that was once the winter coat of Muk-wa, the bear, and Iagoo told them the story of How the Summer Came:

In the wild forest that borders the Great Lake, there once lived a mighty hunter named O-jeeg. No one knew the woods so well as he; where others would be lost without a trail to guide them, he found his way easily and quickly by day or night, through the trackless tangle of trees and underbrush. Where the red deer fled, he followed; the bear could not escape his swift pursuit. He had the cunning of the fox, the endurance of the wolf, the speed of the wild turkey when it runs at the scent of danger.

When O-jeeg shot an arrow, it always hit the mark. When he set out on a journey, no storm or snow could turn him back. He did everything he said he would do and did it well.

Thus, it was that some men came to believe that O-jeeg was a Manito, the Indian name for one who has magic powers. This much was certain; whenever O-jeeg wished to do so, he could change himself into the little animal known as the Fisher or Marten.

Perhaps that is why he was on such friendly terms with some of the animals, who were always willing to help him when he called upon them. Among these were the otter, the beaver, the lynx, the badger, and the wolverine. There came a time, as we shall see, when he needed their services badly and they were not slow in coming to his assistance.

O-jeeg had a wife whom he dearly loved and a son of 13 years who promised to be as great a hunter as his father. Already, he had shown great skill with the bow and arrow; if some accident should prevent O-jeeg from supplying the family with the game upon which they lived, his son felt sure that he himself could shoot as many squirrels and turkeys as they needed to keep from starving. With O-jeeg to bring them venison, bear's meat, and wild turkey, they had thus far plenty to eat. Had it not been for the cold, the boy would have been happy enough. They had warm clothing made from deerskin and furs; to keep their fire burning, they had all the wood in the forest. Yet, in spite of this, the cold was a great trial, for it was always winter and the deep snow never melted.

Some wise old men had somewhere heard that the sky was not only the roof of our own world but also was the floor of a beautiful world beyond, a land where birds with bright feathers sang sweetly through a pleasant warm season called Summer.

It was a pretty story that people wished to believe; and likely enough, they said, when you came to think that the sun was so far away from the earth and so close to the sky itself.

The boy used to dream about it and wonder what could be done. His father could do anything; some men said he was a Manito. Perhaps he could find some way to bring Summer to earth. That would be the greatest thing of all.

Sometimes it was so cold that when the boy went into the woods his fingers would be frostbitten, and then he could not fit the notch of his arrow to the bowstring and was obliged to go back home without any game whatsoever. One day he had wandered far in the forest and was returning empty-handed when he saw a red squirrel seated on its hind legs on the stump of a tree. The squirrel was gnawing a pine cone and did not try to run away when the young hunter came near, and then the little animal spoke:

"My grandson," said he, "there is something I wish to tell you that you will be pleased to hear. Put away your arrows and do not try to shoot me, and I shall give you some good advice."

The boy was surprised, but he unstrung his bow and put the arrow in his quiver.

"Now," said the squirrel, "listen carefully to what I have to say. The earth is always covered with snow, and the frost bites your fingers and makes you unhappy. I dislike the cold as much as you do. To tell the truth, there is little enough for me to eat in these woods with the ground frozen hard all the time. You can see how thin I am, for there is not much fat in a pine cone. If someone could manage to bring the Summer down from the sky, it would be a great blessing."

"Is it really true, then," asked the boy, "that up beyond the sky is a pleasant warm land where Winter only stays a few moons?"

"Yes, it is true," said the squirrel. "We animals have known it for a long time. Ken-eu, the war-eagle who soars near the sun, once saw a small crack in the sky. The crack was made by Way-wass-i-mo, the Lightning, in a great storm that covered all the earth with water. Ken-eu, the war-eagle, felt the warm air leaking through; but the people who live up above mended the crack the very next moment, and the sky has never leaked again."

"Then our wise old men were right," said the boy. "O-jeeg, my father, can do most anything he has a mind to. Do you suppose if he tried hard enough, he could get through the sky and bring the Summer down to us?"

"Of course!" exclaimed the squirrel. "That is why I spoke to you about it. Your father is a Manito. If you beg him hard enough and tell him how unhappy you are, he is sure to make the attempt. When you go back, show him your frostbitten fingers. Tell him how you tramp all day through the snow and how difficult it is to make your way home. Tell him that someday you may be frozen stiff and never get back at all, and then he will do as you ask because he loves you very much."

The boy thanked the squirrel and promised to follow this advice. From that day, he gave his father no peace.

At last, O-jeeg said to him, "My son, what you ask me to do is a dangerous thing, and I do not know what may come of it; but my power as a Manito was given me for a good purpose, and I can put it to no better use than to try to bring the Summer down from the sky and make the world a more pleasant place to live in."

He then prepared a feast to which he invited his friends, the otter, the beaver, the lynx, the badger, and the wolverine; and they all put their heads together to decide what was best to be done. The lynx was the first to speak. He had traveled far on his long legs and had been to many strange places. Besides, if you had good strong eyes and you looked at the sky on a clear night when there was no moon, you could see a little group of stars that the wise old men said was exactly like a lynx. It gave him a certain importance, especially in matters of this kind; so, when he began to speak, the others listened with great respect.

"There is a high mountain," said he, "that none of you has ever seen. No one ever saw the top because it is always hidden by the clouds, but I am told it is the highest mountain in the world and almost touches the sky."

The otter began to laugh. He is the only animal that can do this; sometimes he laughs for no particular reason unless it is that he thinks himself more clever than the other animals and likes to "show off."

"What are you laughing at?" asked the lynx.

"Oh, nothing," answered the otter. "I was just laughing."

"It will get you into trouble someday," said the lynx. "Just because you never heard of this mountain, you think it is not there."

"Do you know how to get to it?" asked O-jeeg. "If we could climb to the top, we might find a way to break through the sky. It seems a good plan."

"That is what I was thinking," said the lynx. "It is true I don't know just where it is; but a moon's journey from here, there lives a Manito who has the shape of a giant. He knows, and he could tell us."

So, O-jeeg bade good-bye to his wife and his little son; and the next day, the lynx began the long journey, with O-jeeg and the others following close behind. It was just as the lynx has said. When they had traveled day and night for a moon, they came to a lodge, as the white men call an Indian's tent, and there was the Manito standing in the doorway. He was a queer-looking man such as they had never before seen, with an enormous head and three eyes, one eye being set in his forehead above the other two.

He invited them into the lodge and set some meat before them; but he had such an odd look, and his movements were so awkward, that the otter could not help laughing. At this, the eye in the Manito's forehead grew red like a live coal; and he made a leap at the otter, who barely managed to slip through the doorway out into the bitter cold and darkness of the night, without having tasted a morsel of supper.

When the otter had gone, the Manito seemed satisfied and told them they could spend the night in his lodge. They did so; and O-jeeg, who stayed awake while his friends slept, noticed that only two of the Manito's eyes were closed while the one in his forehead remained wide open.

In the morning, the Manito told O-jeeg to travel straight toward the North Star and that in 20 suns (the Indian name for days) they would reach the mountain.

"As you are a Manito yourself," he said, "you may be able to climb to the top and to take your friends with you, but I cannot promise that you will be able to get down again."

"If it is close enough to the sky," answered O-jeeg, "that is all I ask."

Once more, they set out. On their way, they met the otter, who laughed again when he saw them; but this time he laughed because he was glad to find them and glad to get some meat that O-jeeg had saved from the Manito's supper.

In 20 days, they came to the foot of a mountain. Then up and up they climbed until they passed quite through the clouds and up once more until at last they stopped, all out of breath, and sat down to rest on the highest peak in the world. To their great delight, the sky seemed so close that they could almost touch it.

O-jeeg and his comrades filled their pipes; but before smoking, they called out to the Great Spirit, asking for success in their attempt. In Indian fashion, they pointed to the earth, to the sky overhead, and to the four winds.

"Now," said O-jeeg when they had finished smoking, "which of you can jump the highest?"

The otter grinned.

"Jump, then!" commanded O-jeeg.

The otter jumped and, sure enough, his head hit the sky; but the sky was the harder of the two, and back he fell. When he struck the ground, he began to slide down the mountain; soon he was out of sight, and they saw him no more.

"Ugh!" grunted the lynx. "He is laughing on the other side of his mouth."

It was the beaver's turn. He, too, hit the sky but fell down in a heap. The badger and the lynx and had no better luck, and their heads ached for a long time afterward.

"It all depends on you," said O-jeeg to the wolverine. "You are the strongest of them all. Ready, now...jump!"

The wolverine jumped and fell but came down on his feet, sound and whole.

"Good!" cried O-jeeg. "Try again!"

This time the wolverine made a dent in the sky.

"It's cracking!" exclaimed O-jeeg. "Now, once more!"

For the third time, the wolverine jumped. Through the sky he went, passing out of sight, and O-jeeg quickly followed him.

Looking around them, they beheld a beautiful land. O-jeeg, who had spent his life among the snows, stood like a man who dreams, wondering if it could be true. He had left behind him a bare world white with winter, whose waters were always frozen, a world without song or color. He had now come into a country that was a great green plain with flowers of many hues, where birds of bright plumage sang amid the leafy branches of trees hung with golden fruit. Streams wandered through the meadows and flowed into lovely lakes. The air was mild and filled with the perfume from a million blossoms. It was Summer.

Along the banks of a lake were the lodges in which lived the people of the sky, who could be seen some distance away. The lodges were empty but before them were hung cages in which there were many beautiful birds. Already, the warm air of Summer had begun to rush through the hole made by the wolverine; and O-jeeg now made haste to open the cages, so that the birds could follow.

The sky-dwellers saw what was happening and raised a great shout; but Spring, Summer, and Autumn had already escaped through the opening into the world below and many of the birds, as well.

The wolverine, too, had managed to reach the hole and descend to earth before the sky-dwellers could catch him, but O-jeeg was not so fortunate. There were still some birds remaining that he knew his son would like to see, so he went on opening the cages. By this time, the sky-dwellers had closed the hole, and O-jeeg was too late.

As the sky-dwellers pursued him, he changed himself into the Fisher and ran along the plain toward the North at the top of his speed. In the form of the Fisher, he could run faster. Also, when he took this shape, no arrow could injure him unless it hit a spot near the tip of his tail; but the sky-dwellers ran even faster, and the Father climbed a tall tree. They were good marksmen, and they shot a great many arrows until at last one of these chanced to hit the fatal spot; and then the Fisher knew that his time had come.

Now, he saw that some of his enemies were marked with the totems, or family arms, of his own tribe. "My Cousins!" he called to them. "I beg of you that you go away and leave me here alone."

The sky-dwellers granted his request. When they had gone, the Fisher came down from the tree and wandered around for a time, seeking some opening in the plain through which he might return to the earth; but there was no opening, so at last, feeling weak and faint, he stretched himself flat on the floor of the sky through which the stars may be seen from the world below.

"I have kept my promise," he said with a sigh of content. "My son will now enjoy the summer and so will all the people who dwell on the earth.

Through the ages to come, I shall be set as a sign in the heavens; and my name will be spoken with praise. I am satisfied."

So, it came about that the Fisher remained in the sky, where you can see him plainly for yourself on a clear night with the arrow through his tail. The Indians call them the Fisher Stars, O-jeeg An-nung, but to white men are they known as the constellation of the Plough.

Larned W.T.L. *How Summer Came.* American Indian Fairy Tales, Story Number 6, P. F. Volland Company, Chicago, Illinois, 1921.

JUNE

What is so rare as a day in June?

Then, if ever, come the perfect days;

Then heaven tries earth if it be a tune,

and over it softly her warm ear lays:

Whether we look, or whether we listen,

We hear life murmur, or see it glisten.

— JAMES RUSSELL LOWELL

Strawberry Moon — Heart-Berry Moon

"There is a time for everything, and a season for every activity under heaven."
                                              – Ecclesiastes 3:1

Strawberry photo
courtesy of Jerie Artz

The time of the Strawberry Moon was of great significance for American Indian people. They believed the strawberry plant was a sacred gift from their Creator. The fruit was used to symbolize his generosity in their Strawberry Thanksgiving Ceremony. Strawberries were one of the first fruits to ripen in early summer, so after the Planting Moon the women would begin collecting the berries to prepare for the ceremony. It was one of several thanksgiving ceremonies held throughout the lunar year to express gratitude for their bounteous harvests, their families and traditions.

Strawberries ripen in early summer across Northeastern regions of the United States. The berries were an important seasonal food source; they were eaten fresh, made into special drinks and added to soups and breads. The fruit was mashed and made into small cakes that were dried for later use. All parts of the plants were harvested with careful attention paid to gathering time. The women knew the medicinal virtue of each part of the plant changed throughout the growing season. They used an infusion of the whole plant to regulate menses. The young leaves were used to make a tea-like beverage. They kept products made from the strawberry plant in their homes all year to lift their spirits and ensure happiness.

The berry's heart shape led to the name Heart-Berry Moon and the fruit is associated with forgiveness in American Indian culture. Some American Indian tribes believed that strawberries grew along the path to paradise.

One of many American Indian legends passed down through the centuries describes the origin of the strawberry. Man and woman lived together happily until one day they began to quarrel. The woman became so angry she decided to leave the man and find another place to dwell. The man soon began to miss her and wept with sorrow over their broken friendship. The Great Spirit heard the sorrowful noise and looked down with pity upon the weeping man. "Would you like to be reunited with the woman?" he asked. The man said yes, and the Great Spirit told him to go find her. When the man saw her walking ahead of him, he called out to her. She was still angry and would not stop, nor did she notice the blueberry bushes placed along the path by the Great Spirit. She just kept walking. The Great Spirit caused juicy blackberries with thorns to grow in her way, yet she still wouldn't stop, even though the thorns scratched her and tore her clothes. Finally the Great Spirit created a creeping green plant with fragrant heart-shaped berries and placed it by her feet. The berry's bright color and delightful scent was so entrancing she stopped and picked one and ate it. She stayed and gathered more to eat. When she saw the man coming to apologize, they made up and shared the berries. The strawberry symbolizes the love of the first man and woman. Native people named it the heart-berry.

Strawberries belong to the rose family and they share many of the same family characteristics. Both plants are useful as well as beautiful. The most common wild strawberry found in the Eastern and Central part of the United States is the Virginia Strawberry *(Fragaria virginiana)*, a perennial plant that grows in fields and dry woodland and forest openings. Some cultivated strawberries you find in stores are hybrids developed from this native species. The wild plant is believed to be pollinated by native bees — the carpenter, leaf cutting, digger bee and the bumblebee. It is a larval host plant and nectar source for the Gray Hairstreak butterfly. People and birds both love the berries. Strawberry festivals are held every summer throughout the United States in honor of this berry.

O, my love is like a red, red rose,
That's newly sprung in June.
– ROBERT BURNS

The festival is a wonderful tradition that has its origins in the first grateful thanks of American Indian peoples.

Chippewa Customs by Frances Densmore is an excellent resource book for a more in-depth account of North American Indian life. She describes in great detail their customs, seasonal activities and the role of women in American Indian culture.

Summer

Summer solstice occurs on June 21 or 22, marking the official beginning of the summer season in the Northern Hemisphere. It is the longest day of the year. The days grow shorter with each passing day until we reach the fall equinox, when the days and nights are equal again. In ancient days, many countries in Europe called June the *"Honey Moon"* because it was the best time to harvest honey from bees. A mead drink made from fermented honey was served at wedding ceremonies performed at summer solstice. Many early cultures celebrated summer solstice with huge community bonfires or by rolling a burning wheel down a hill, imitating the sun's course in the sky as its annual declination begins.

Legend has it that herbs harvested during summer solstice have special powers for healing and protection. It was thought to be a magical night and you would be able to see fairies and elves if you picked fern spores at the stroke of midnight. Herbs associated with midsummer traditions are chamomile, cinquefoil, elder, fennel, roses, mugwort, rue and St. John's Wort.

American Indian children loved playing sports and games and Frances Densmore describes many of their favorite seasonal activities in her book, *Chippewa Customs.* They were very creative at adapting stones, sticks and plants for play and sport. The American Indians believed the

How softly runs
the afternoon,
Beneath the billowy
clouds of June.
– CHARLES H. TOWNE

butterfly represented the spirit of childish play and images of butterflies were used to decorate children's belongings. Children were taught to invite butterflies to join in their games. The butterfly game (hide and seek) was one such game. Children would gather in a group and draw lots from a specially prepared bundle of sticks of different lengths tied together with basswood fibers. The one who drew the longest stick covered his eyes as the other children ran and hid. After everyone found a place to hide, he would search for them. While he was searching he would chant the butterfly song "Me-e-mem-gwe, me-e-mem-gwe (butterfly) show me where to go."

Butterfly Gardens
By Nancy Ewing, Michigan Master Gardener

Why create a butterfly garden? Because most of their natural habitats, (like those of so many other species of birds, insects and animals) are being destroyed at alarming rates, and creating a butterfly garden is an environmentally friendly thing to do. Where we live is where they once lived. While we cannot restore many other native habitats on our own, we can for butterflies, and can do it a way that is pleasing to us.

Tiger Swallowtail on Butterfly Bush

How? By planting a garden! What makes this even easier is that many of the plants commonly found in our gardens are the same plants that attract butterflies. You don't need a large garden to bring in the butterflies. One well-placed butterfly bush will bring amazing results, but the more nectar-producing plants that you have, the more butterflies you will have. Clumping these nectar-producing plants together in one area is the best way to attract butterflies. It takes a great deal of energy to fly from flower to flower, so putting the flowers closer together creates a more appealing butterfly diner.

You will want plants of varying heights. The larger the butterfly, the taller the height of the flowers that they prefer to feed to upon. You will also want to plant a variety of plants, with varying bloom times, so there is a continual food source throughout the season. Annuals are very helpful in fulfilling this need.

Now that you have attracted butterflies, you want to keep them. You do this by providing caterpillar foods or larval host plants. This is the more difficult task. First, observe your garden and identify what types of butterflies you have visiting. Then you can plant the larval host plants. Many species of butterfly lay their eggs on only one or two varieties of plants, making this task even more difficult. Many larval host plants are wildflowers, weeds and grasses that belong in an informal setting, not in a flower garden. So you want to designate a separate area for these. Start with one or two species of butterflies that you want to create a habitat for. Monarch is an obvious choice for our area. They are plentiful and their host plants, the milkweeds and butterfly weeds, are readily available and can be incorporated into the flower garden. For more resources and information go to Butterflies and Moths of North America, www.butterfliesandmoths.org.

Below are listed some essential elements to keep butterflies in your garden:

Monarch Butterfly

Sunshine

Butterflies are cold-blooded and need the sun to warm them. For this reason your garden should be planted in a warm, sunny area. A few large stones to capture heat and provide basking spots are good additions.

Shelter From Winds

Butterflies rarely fly during strong winds and cannot feed. So plant your garden where the wind will be blocked by a building or taller shrubs and trees.

Water and Minerals

Butterflies need to drink, which is how they get minerals. They often drink from muddy puddles. You can make a 'sand box' for butterflies by burying a shallow water holding container and covering it with soil. The container will hold moisture, giving the butterflies a place to drink.

A Pesticide-Free Environment

Although you may find chemical pesticides an easy way to get rid of unwanted insects and weeds, they also prevent friendly insects, like butterflies, from making your garden their home. Try alternatives to chemical pesticides, such as spot treating with insecticidal soaps.

Nectar Sources

Nectar is a sugar-rich substance that is required for energy used in flight. Plants that attract butterflies are those that are both sun loving and nectar producing. It is recommended that you plant similar intense colors in masses, such as yellows, dark pinks, purples, reds and oranges. Plant a variety of plants that will provide blooms all season long, adding some annuals for continuous flowers. Because butterflies need to rest while feeding, they like plants with large flower heads that give them room to land and feed simultaneously.

Host Plants

You will also need to provide the type of plants that butterflies require for their larvae (caterpillars) to eat. These vary by butterfly species, and are the 'tricky' part of butterfly gardening. Mother Nature is very skilled at multi-tasking; many of the plants, trees and shrubs planted for butterflies will also provide shelter, nesting sites or berries for birds.

Larval Host Plants Listed by Butterfly Species

This is not a complete list, but a solid guide for starting and establishing a butterfly garden.

Butterfly	Host Plant
Black Swallowtail	Parsley, Wild Carrot (Queen Ann's Lace), Dill, Parsnip, Fennel
Spicebush Swallowtail	Spice Bush, Sassafras
Giant Swallowtail	Prickly Ash, Hop Trees
Pipe-vine Swallowtail	Dutchman's Pipe-vine
Tiger Swallowtail	Wild Black Cherry, Ash, Chokecherry, Lilac, Tulip Tree
Great Spangled Fritillary	Violets
Painted Lady	Thistles
Red Admiral	Nettles, False Nettle
Viceroy	Willows, Poplars, Plums, Cherries
Monarch	Milkweeds
Sulphurs	White Sweet Clover, Legumes, Crown Vetch
Coppers	Sorrel, Docks
Eastern Tailed Blue	Pea family
Pearl Crescent	Asters
Milbert's Tortoise Shell	Nettles
American Painted Lady	Pearly Everlasting, Pussy-toes, Burdock
Painted Lady	Thistle, Burdock, Aster, Hollyhock, Mallow
Little Wood Satyr	Grasses
Skippers	Grasses
Buckeye	Snapdragon, Turtlehead, Verbena
Baltimore	Turtlehead
Mourning Cloak	Willows, Elms, Aspens

Nectar Plants for Butterflies

Shrubs

Azalea

Cotoneaster

Daphne

Deutzia

Lilac

Privets

Rose of Sharon

Spicebush

Sumac

Viburnum

Annuals

Ageratum

Alyssum

Candytuft

Cleome

Cosmos

Heliotrope

Hibiscus

Hollyhock

Lantana

Marigold

Nicotiana

Penta

Petunia

Verbena

Zinnia

Perennials

Asters

Bee Balm (Monarda)

Black-Eyed Susan (Rudbeckia)

Boneset

Butterfly Weed

Coral bells

Coreopsis

Daisy

Daylily

Gaillardia

Globe Thistle

Goldenrod

Honeysuckle

Joe-Pye Weed

Lavender

Liatris

Milkweed

Phlox

Primrose

Purple Coneflower (Echinacea)

Red Valerian

Scabiosa

Sedum

Swamp Milkweed

Veronica

Yarrow

Common Milkweed (*Asclepias syriaca*)

Common Milkweed is the best known of the 100 or so milkweed species native to North America. It occurs naturally in prairies, pastures, fields and along roadsides across the Eastern United States.

American Indians used milkweed for fiber, food and medicine. They taught early European settlers how to safely prepare the plant's young shoots as a seasonal vegetable (a practice not recommended today). The American Indian's relationship to the plant world was the product of centuries of accumulated knowledge; they were fully acquainted with the plant's characteristics in all stages of growth and at various times of the year. The roots were used medicinally and the milk of the fresh plant was used to remove warts. The flower buds were added to soups and dried buds were stored for winter use. Milkweed's tough fibers were used for making cords, ropes, weaving cloth and popgun wadding by little boys.

Following the life cycle of the common milkweed plant can be a fun summer adventure. It's a very interesting plant readily available for observation and serves many purposes in the natural environment. Plants can provide clues about pollution in our air, land and water. Milkweed is very sensitive to ozone and can be used to detect air pollution. Ozone injury on milkweed can be seen on the leaves.

Milkweed is named for the poisonous, sticky, milky substance which exudes from its leaves when broken or pierced. The plant is toxic to many insects and animals, but the few who have developed a resistance owe their survival to it. It is the host plant for the Monarch butterfly. They lay eggs on the milkweed leaves and after the eggs hatch the baby caterpillars eat the leaves. The poison they ingest doesn't hurt the caterpillars, but it makes them taste bad to birds and other predators. If you find a patch of milkweed in late June you can watch this remarkable butterfly's life cycle from start to finish.

To make a wish come true, whisper it to a butterfly. Upon these wings it will be taken to heaven and granted. For they are the messengers of the Great Spirit.

– NATIVE AMERICAN LEGEND

When observing the butterflies you will notice brightly-colored milkweed bugs and milkweed beetles feasting on the leaves. A milkweed plant can be a very active place on a warm June afternoon.

Monarchs are noted for their exceptional winter migration to Mexico. This migration is so long it takes three generations to complete. Therefore, the Monarchs that return to us each spring are not the same ones that left us last fall, but their children's grandchildren. They depend on milkweed and other food sources traveling back and forth. With the mowing and herbicide use along roadways and disappearing green spaces, they are having trouble finding food and shelter during their annual fall and spring migrations.

We need to ensure that future generations of children and Monarch butterflies have safe places to meet and play. Everyone can help by letting milkweed colonies grow in wild waysides and fields. Planting milkweed in out of the way places in home landscapes, schools and parks will create safe havens for these special creatures. We can all make a difference, one backyard or neighborhood at a time. For help getting started go to www.monarchwatch.org.

Nocturnal Adventures

A summer evening is a great time to be outdoors. No matter how familiar you think you are with your backyard and beyond, it becomes a totally different world as day shifts to night. Flickering fireflies, chirping crickets, drumming cicadas, floating moths and swooping bats are all part of the summer night shift.

Most of us are familiar with diurnal animals, those that are active during the day, and nocturnal animals, those that are active at night. The twilight zone is a magical time of day that frequently passes unnoticed in the hustle and bustle of everyday living. Twilight is the time proceeding

And tis my faith that every flower. Enjoys the air it breaths.
— WILLIAM WORDSWORTH

Treat the Earth well. It was not given to you by your parents, it was loaned to you by your children. We do not inherit the Earth from our Ancestors, we borrow it from our children.
—ANCIENT INDIAN PROVERB

sunrise and again following sunset when there is still a bit of light outside but the sun is not in the sky. Crepuscular animals are active at dawn and dusk.

There is so much to observe while you watch the sun set on a clear summer evening. Our sense of sight becomes diminished in the dark, but our sense of smell, sound and touch becomes more acute as the temperature cools and the air becomes more humid as day changes to night. After the sun falls below the eastern horizon and the moon appears, the world becomes black, white and gray.

Bats emerge at twilight, sweeping and swirling in search of night-flying insects. A single little brown bat can catch more than 600 mosquitoes in an hour.

Fireflies, or lightning bugs kick off the summer season with scattered appearances in early June and by mid-June are out in full force. There are believed to be 20 to 30 species of fireflies in the eastern half of the United States. Not all fireflies flash, but those that do have their own characteristic method of flashing, distinguishable by such features as flight levels, color, number of flashes and intervals between flashes and the time of night or year they are active. Fireflies are the only insects that flash on and off in distinct signals. Because they are only active at night, the flashing is a visual way of attracting a mate. Observing fireflies is a favorite summer activity that dates back to pre-colonial times in America. They can be seen flashing over meadows, lawns and at the edges of woodland areas. As you watch fireflies you can look for the different signal patterns and colors of light. Firefly Watch, www.mos.org/fireflywatch, is an interesting Citizen Science Project sponsored by the Museum of Science in Boston, Massachusetts. They are looking for volunteers to collect data about fireflies across the United States. The site is full of great tips for families or neighborhood organizations who would like to learn how to provide safe habitats for these enchanting little beetles.

Night-blooming flowers are usually light pink, white or a creamy beige and generate a stronger fragrance at night, a strategy meant to attract night-pollinating insects. Moths and butterflies share some similarities; however, there are some distinct features you can observe to help distinguish whether you are looking at a moth or butterfly. Moths are mostly active at night, have duller colors and feathery antennae without a club. Their job on nature's night shift is to help pollinate the light-colored blooming flowers. On the other hand, butterflies prefer red, green and yellow colors which do not show up well in the dark. A moth's feathery antennae act like a nose responding to the scent molecules in the air. They are attracted to the fragrant night-blooming flowers in the garden. You can make a moth bait to encourage more moths to stop by for an evening visit. Mash up a rotten banana with some sugar and fruit juice and let it set for a few hours. Just before dusk brush the mixture on several trees and then watch to see who visits after dark.

CHANT TO A FIREFLY
CHIPPEWA POEM

Flitting-white-fire-insect! Waving-white-fire-bug give me light before I go to bed! Give me light before I go to sleep. Come, little dancing white-fire-bug! Come, little flitting white-fire-beast! Light me with your bright white-flame-instrument – your little candle.

Literal translation
Collected by
Henry Rowe Schoolcraft

GOOD FRIENDS — OUR NATIVE TREES

Canadian Serviceberry *(Amelanchier canadensis)*

The Canadian Serviceberry is a delightful small tree or shrub that occurs naturally throughout the Eastern United States, southward to Florida and westward to Minnesota. In early spring, its shaggy white flowers blooming on upsweeping branches are sometimes mistaken for Flowering Dogwood. Both trees have been used as season indicators by American Indians for thousands of years. Early colonists who came to America and settled in wilderness areas far from civilization would wait until spring to hold memorial services for loved ones who passed away during the winter. When colonists saw serviceberry trees blooming in

the spring woodlands and roadside thickets, they knew the preacher would soon arrive to hold "services" for their loved ones. Serviceberry is one of several names attached to the tree. It is also called Juneberry because its fruit ripens in June. The Lenni Lenape people along the East Coast called the tree shadbush, because when they saw the blossoms they knew the shad fish would be coming upstream to spawn. The United States National Phenology Network, http://www.usanpn.org/participate, recommends this tree to their citizen science partners for observation and data collection.

American Indians used an infusion of serviceberry bark for a bath for children with worms and a decoction of the inner bark was used as a disinfectant. The tasty tiny fruit was mashed and made into little cakes and dried for winter use. The tree's wood was used to make arrows.

Serviceberry is a very attractive small deciduous tree that can fill several niches in a home landscaping scheme. It has charming, dainty, white spring flowers, edible fruit that both humans and wildlife find tasty and eye-catching orange fall foliage. The tree tolerates most soils and is very drought tolerant once established. It can be planted individually or in group plantings and hedges. It is a fast growing tree that can quickly reach heights of 15 to 20 feet. Female native bees collect pollen and nectar from the shaggy white flowers of the serviceberry in early spring to make wee loafs of "bee bread" to feed newly hatching baby bees. Birds nesting in the branches love the berries and eat them as fast as they ripen. Orioles, robins, catbirds, and Eastern Bluebirds all like to eat the berries. If you can beat the birds to them, there are many recipes for preserving the berries for jams and compotes.

Canadian Serviceberry

ACTIVITIES AND ADVENTURES FROM
THE BACKYARD AND BEYOND

Make a butterfly puddle-stone!

The following information is used with permission from Nancy Ewing, Michigan Master Gardener. What is a butterfly puddle-stone? A puddle-stone is a place you create which gives butterflies a place to obtain the moisture and nutrients that they need that are not provided through nectar. A puddle-stone can be made from any object placed on the ground that will hold soil and retain moisture. It can be as simple as a plate or you can get creative! I have used mirrors which reflect light and attract butterflies, decorative plates, leaf molds (directions follow) or coffee can lids.

The basic premise is to fill the object with ordinary road gravel, which is a combination of sand, gravel and road salts. Keep the soil moist and watch for the butterflies to come. They will insert their proboscis into the damp soil and withdraw moisture and minerals. This behavior is primarily seen from male butterflies and is considered to be a social behavior as well as providing the necessary nutrients for breeding.

Butterfly Puddle-Stone

To make a leaf mold, you will need a large, deeply-veined leaf of your choice. Hosta, rhubarb, oak leaf hydrangea and water lily are all good choices. The leaf is best if fresh and in good condition. After gathering your materials and finding the perfect leaf or leaves, you can begin.

Materials needed:
- Latex concrete patch mix
- Clean play sand
- Rubber gloves
- Mixing tray or wheel barrow
- Water
- Plastic sheeting
- Flat working surface (someplace you can let the mold sit for a few days to cure)

Lay a sheet of plastic that is at least 12 inches larger than your leaf on your work surface. Cover the center of the plastic with play sand, building it up to a mound in the middle. Lay your leaf face down on the sand (veins facing upward).

Mix the dry latex concrete patch blend in your tray or wheelbarrow with just enough water to give it the texture of set pudding. With gloves on, beginning in the middle of the leaf, place handfuls of the concrete mixture directly on top of the leaf, working it out just to the leaf edge. You want the mold to be at least one-inch thick.

Cover the entire project with another sheet of plastic and let it sit undisturbed for a few days. A cool dry location is best. Remove the plastic and carefully lift the leaf and concrete from the sand. The leaf will peel away from the concrete. Sometimes it takes a day or two for this to happen, and the leaf may need to dry out first. You may gently use a tool to remove the last of the leaf, if necessary.

Wash the mold with water and let cure for a few days before placing in the garden for use.

Build a Toad Abode

Keep an eye out for tiny toadlets hopping about your backyard and beyond. It's hard to believe these tiny brown, bumpy amphibians were slimy black tadpoles swimming in area ponds just two months ago. Very few of these dime-size babies survive to adulthood. Give them a helping hand by providing a safe place to live and grow up.

Collect three flat-sided rocks for the walls and one larger rock to lay on top for the roof. Find a pleasant shady spot in your yard or garden beneath a leafy shrub or plant. Lay the three flat-sided walls on the ground facing away from the sun. Lay the larger flat stone across the top of the walls to create a roof. Push some soil up around the walls to help hold the rocks in place.

The tiny toad can comfortably rest in his wee cave during the day and venture out in the evening. Frogs, toads and salamanders prefer the humid, cool night air and like to move about after dark.

Plant a Strawberry Jar

Strawberry jars are urn-shaped pots with random planting pockets on their sides. While they were created especially for strawberries, herbs and succulents also grow well in them. Due to their unusual shape, you need to add a special drainage column when planting to ensure even watering. Strawberry jars are available in various sizes and shapes at your local garden center. Even though they need to be watered more frequently, plain or glazed terra cotta jars make the most attractive containers.

There are a wide variety of strawberry plants to choose from. It is most important to look for a strawberry plant that produces fewer runners. Alpine Strawberry is a variety that grows well in containers.

Materials needed:
- High quality potting soil mix and compost
- Strawberry jar
- Water
- Coffee filter
- Paper towel roll
- Pea gravel
- Bare root strawberry plants

Mix together 2/3 potting soil with 1/3 compost in a container and slowly add water until the mix is moist. Let set.

Lay the coffee filter in the bottom of the pot to cover the drainage hole. Layer an inch or so of soil over the bottom covering the coffee filter. Stand the paper towel roll up in the center of the pot and fill the roll with pea gravel.

Add soil up to the first level of pockets, keeping the paper roll centered as best you can. Carefully push one plant into each pocket leaving the crown of the plant exposed. (The point where the stem and roots meet.) Firm the plants in place and add soil mixture up to the rims of the next set of pockets. Plant these pockets in the same manner as the first set and repeat until you get a few inches below the top of the container.

Plant three strawberry plants in a triangle and fill in around them with soil. Water slowly so the water filters to the bottom and through the pockets. Set in a shady spot for a few days to let the plants settle and then place in a sunny location. Pinch off the first flowers that bloom so the plants can focus on establishing strong roots.

Bug Trap

It's hard to see everything crawling around on the ground at night. But there is a way to check out some of the bugs who have been boogieing about your yard in the dark. Catch some!

Materials needed:
- Spade
- Wide-mouth jar
- Dried bits of fruit, chopped banana peel mixed with some dirt
- Large leaf or piece of wood or bark

Dig a hole and sink the jar down until the top of the jar is even with the ground. Drop bits of dried fruit and chopped banana peel into the bottom of the jar. Cover the top of the jar with a piece of wood, bark or large leaf propped with a stick or rock to keep out water. Leave the jar overnight and in the morning check to see what is in the jar. Always practice catch and release. Observe, identify and then set free.

THYMELY TIPS AND SAGE ADVICE

• Make sure everyone takes safety precautions and properly prepares for outdoor nocturnal adventures. Ticks and mosquitoes are active at night so be sure to put on some bug spray. If you are going to be in tall grass or woods, wear a long-sleeve shirt and long pants. Tuck your shirt into your pants and your pant legs into your socks and spray your shoes, socks and pant legs with insect repellent. When your adventures are finished, be sure to check yourself for ticks.

• A flashlight is a handy tool to have when conducting nocturnal adventures. If you cover the face of the flashlight with blue or red acetate or cellophane, the light will not be as easily detected by the animals and insects active after dark.

• *"Leaves of three; beware of me."* Watch out for poison ivy. Learn to properly identify poison ivy and poison oak. The Poison Ivy, Oak and Sumac Information Center Web site will help you identify the plants, www.poisonivy.us.

• Mulch newly-planted woody plants to conserve moisture, moderate soil temperatures and reduce weed growth. The proper choice and application of mulch can noticeably improve the overall health and vitality of trees and other landscape plants.

• Stake your top-heavy herbaceous plants early this month. It will keep them protected from wind and storm damage and help preserve their symmetry and beauty throughout the growing season.

• Last call: early summer is the time to sow some extra seeds of beets, carrots, endive, cauliflower, parsnips and radishes for fall and early winter harvesting.

JULY

Hot July brings cooling showers.

— SARAH COOLIDGE

Buck Moon — Thunder Moon — Bee Moon

White-tailed deer are native to North America from Canada to Texas. They can easily be observed in a variety of habitats — woodland, urban and agricultural. The deer is so named for the white underside of its distinctive tail, which is raised in order to warn other deer of potential danger or when mother does flag their fawns to follow them.

White-tailed deer were an important source of food for American Indian cultures. They taught early pioneers to make practical and effective use of every scrap of the meat, bone, and hide. Deer hides were used for clothing and blankets, tools were crafted from the bones and antlers. Male deer shed their antlers every year in late fall after mating and begin growing a new pair the following summer. Some American Indian tribes referred to July's full moon as the Full Buck Moon because it was the month when the bucks begin to re-grow their antlers.

Before Europeans settled in America, deer herd densities averaged six to ten deer per square mile. Today some suburban deer herd densities are approaching twenty or more deer per square mile, drastically changing the flora of cultivated and natural settings and altering vital ecosystems. Cities and towns are working with local citizens and the Department of Natural Resources to find ways to responsively balance this environmental problem.

There are many ancient legends about the origins of thunder. American Indians described this natural force of nature as the thunderbird; the one who brings the rain. The thunderbird motif has been used for centuries in American Indian art work. It is also a prominent figure in Native North American mythology, depicted as a powerful spirit in the form of a bird. Lightning flashes from his beak, and the beating of his wings causes the thunder.

Historically, thunderstorms have been a common weather pattern as-

Thonah! Thonah!
There is a voice above
The voice of thunder.
Within the dark cloud.
Again and again
it sounds.
Thonah! Thonah!

Thonah! Thonah!
There is a voice below.
The voice of the
grasshopper.
Among the plants.
Again and again
it sounds.
Thonah! Thonah!

– Navajo
Translated by
Washington Matthews, 1887

sociated with the annual arrival of summer heat and humidity in Eastern and Central regions of the United States. The storms are more frequent and intense in July, so some American Indians called this moon the Thunder Moon

"Avoid the Ash, it counts the flash."

Ancient superstitions have been passed down through the ages concerning the origins of lightning and thunder. Certain plants were associated with attracting or repelling strikes of lightning. The Hazel tree (*Corylus* spp.) is a common species of nut tree native to Europe. The tree was believed to be immune from damage by lightning along with many other wondrous virtues. Farmers would cut and save hazel twigs in early spring. On the first thunderstorm of the season they would fashion crosses from the twigs to lay on top of their grain piles to ensure the grain would keep well into the future. The oak tree and mistletoe were also believed to be lightning plants and were treated with special reverence by ancient peoples. *Euphorbia lathryus* is a familiar species of spurge sold by many U.S. nurseries as a natural mole deterrent. Ancient people living in Europe called this plant Magic Springwort. They coveted the plant because of its reputation of drawing down lightning and dividing a storm.

The Lenni Lenape or Delaware clans living in the Eastern woodlands along the Delaware River referred to this month as the Bee Moon. There are some 4,000 species of native bees in North America. Bumblebees, leaf cutters, orchard mason bees, carpenter, sweat and digger bees are just a few of the natives that were living in America before the arrival of the honeybee.

Honeybees were introduced into America by settlers from England in about 1622. Swarms of honeybees escaped and spread across North America, often ahead of the first settlers. American Indians who encountered honeybees for the first time called them "white man's flies."

When dew is on the grass, Rain will come to pass.
When grass is dry at morning light,
Look for rain before it's night.
– TRADITIONAL LORE

BEE STINGS

If you get stung by a bee, crush up a plantain or bee balm leaf and put it on your skin to reduce the pain and swelling.

Many of our native bees are 50 to a 100 times faster and more efficient at pollinating certain plants than honeybees, plus they are willing to work in cool, rainy weather when honeybees are not active. Other natives are specialist bee species that forage on specific groups of plants, found growing within short distances from their nests.

July is the peak time of activity for insects. They can be seen and heard in gardens, along the edges of forests, meadows and vacant lots. It is also the peak season for wildflowers. They are showier and more vigorous than those of spring with bold, bright blossoms in every imaginable color.

Mother Nature plants this vast variety of green plants and flowers every year to satisfy the needs of the thousands of species of bees, beetles, butterflies, dragonflies and other insects active at this time of year.

June, July and August are the hottest months in the northern hemisphere. Before July ends, summer will reach as far north as it can go, and then will start to slide back toward south and the days will be several minutes shorter every twenty-four hours.

Wasps and bees are normally considered to be pests because of their ability to sting. Bees generally only use their stingers in defense. If you step away slowly when you see them working they'll leave you alone. Only female bees are capable of stinging.

According to the Pollination Protection Institute, 80% of the world's crop plants depend on pollinators. Pollinators, almost all of which are insects, are indispensable partners for an estimated one out of every three mouthfuls of food, spices and condiments we eat and the beverages we drink. They are essential to the plant fibers used in our clothing and many of the medicines that keep us healthy. (Hansen, p. 25) Bees, both managed honeybees and native bees, are the primary pollinators in most areas of North America. Native bee populations are declining due to lack of foraging plants, suitable nesting sites and the use of pesticides.

Bees are fun to watch buzzing about their business. There are some simple steps that families and neighborhood organizations can do to help ensure that bees will always be around to partner with the plants growing in our backyards and beyond. Start by learning to identify the different species of bees, the plants they prefer and their nesting behaviors. Bees are grouped by their nesting habits and by whether they are social or solitary.

Wood-nesting bees are solitary bees. Each female builds and tends her own nest of brood cells. Carpenter bees chew their own nest cells into soft woody substances; other wood-nesting bees build their homes in the tunnels and holes left by beetles and other insects in rotting logs, old stumps and dead and dying trees. They also use twigs and vines of plants with pithy centers such as sumac, wild raspberry, dogwood and elderberry.

Ground-nesting bees require direct access to the soil surface to dig their nests. Each female excavates her own nest tunnel and brood cell, and stocks the cells with nectar and pollen. Although there is just one bee per nest, many of these bees typically nest close to each other. They build their nests in sunny, well-drained sandy areas. Look for small, round holes in the ground surrounded by piles of soil. If you place a glass jar over the hole you can sometimes get a close look as she is leaving her nest to collect nectar and pollen. Be sure to remove the jar after looking at the bee.

Bumblebees are social bees that live in colonies with a queen bee and sterile workers that communicate and cooperate in caring for the young. Bumblebees build their nests in old rodent burrows and small cavities under rocks or wood logs and in grassy areas and meadows.

Bees like a variety of flowers with different shapes, colors and fragrances for foraging. If you have a flower or vegetable garden, plant a

PLANTS BEES LOVE
Daisy
Asters
Sunflower
Mint
Lavender
Rose
Salvia
Bee Balm
Elder
Goldenrod
Basil
Strawberry
Sedum
Raspberry

Ground-Nesting Bees

strip of native wildflowers or "heirloom" open-pollinated plants along the edge of your garden. If you don't have gardens, tuck a few patches of these flowers on the edges of your property. Leave some untidy corners of rough grass and bare patches of earth. Find rotting logs or old stumps and place them in sunny out of the way places for nesting sites.

The Pollinator Partnership Web site, http://pollinator.org, has plant lists and directions for building homemade bees nests. The site includes information for homeowners as well as educators. They offer free materials that children can write for and receive in the mail. The Xerces Society for Invertebrate Conservation also has a very informative Web site for even more positive and productive suggestions to help pollinators, www.xerces.org.

Plant and Insect Weather Forecasters

In times past, people would closely observe the behavior of insects and the opening and closing of wildflowers to predict future weather events. The result of these observations is a wealth of fun weather wisdom passed down through the centuries.

Ants have long been credited with an instinct for predicting seasonal and daily weather events. For instance, if you notice the entrances of ant hills are open, it means clear weather. If they are closed it means a storm is approaching. When they move their dwellings from low ground to high ground, it is a sign of heavy rains. It has also been said that the direction of the prevailing rains of the season can be determined by observing the position of the ant nests, as they position the opening of their nests toward the driest direction. When ants are observed at midsummer, enlarging and building up their dwellings, it is said to be a sign of an early cold winter.

Bees are sensitive to the increased humidity that usually proceeds a

Ant

rain shower and always return to their homes before it rains. The duration of a rain storm can often be foretold by observing the activities of spiders. If they remain on their webs during a rain shower you can be sure the rain will soon be ending; if they disappear from sight the weather will remain unsettled for some time.

When farmers noticed brambles blooming in early June, they would prepare for an early harvest. Chickweed, dandelions, bindweeds and clovers fold their petals before rain showers. When a tree shows the underside of its leaves, wet weather is on the way. These are just a few of the hundreds of interesting bits of weather lore associated with plants and insects. Keeping track of weather events and noting family member's observations will help you sift out which old-time predictions are true or false.

Moon Phases

Early civilizations planted, weeded and harvested according to the phases of the moon and the signs of the zodiac. Many gardeners still believe that planting vegetables during specific phases of the moon will ensure faster seed germination and growth. During the New to Full Moon, when the moon is waxing or growing, they plant above ground crops such as lettuce, peas, tomatoes and peppers. From the Full Moon to the New Moon, as the moon is waning or decreasing, they plant root crops like potatoes, radishes, carrots, beets and turnips.

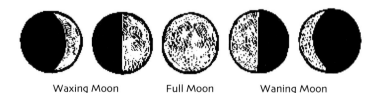

Waxing Moon   Full Moon   Waning Moon

On its 29-day journey around the earth, the moon passes through twelve zodiac signs. Each of these signs are influenced by one of four elemental forces: earth, fire, air or water. Earth and fire are classified as fertile signs associated with the waxing or growing phases of the moon. These are believed to be the optimal times for planting gardens and pruning plants. Weeding, tilling and harvesting tasks are best preformed when the moon is waning and is in a barren sign. For additional information go to www.gardeningbythemoon.com/signs.html.

A Moonlight Garden

Most gardens are generally planned and planted for how they will look and be used during daylight hours. Many families today keep such busy schedules they are often away from home until late evening. By the time they arrive home and have dinner it's dark. Creating a moonlight garden to relax and unwind as a family is a perfect way to spend a summer night. Moon gardens are generally planted with a variety of night blooming white flowers, plants with bright patterns on their foliage and plants with silvery leaves. Moonflowers, members of the morning glory family, do not open until the sun sets. It takes them about ten minutes to open so they are great flowers to show children. The following plants can all be used in a night garden.

PLANTS FOR A MOONLIGHT GARDEN

'Silver Queen,' *Artemisia ludoviciana*

Night Scented Stocks, *Matthiola longipetala*

Evening Primrose, *Oenothera speciosa*

Phlox David, *Phlox paniculata* 'David'

Lamb's Ear, *Stachys byzantina*

Rose of Sharon, *Hibiscus syriaca*, 'Diana'

Variegated Dogwood, *Cornus alba* 'Variegata'

Cosmos, Knee-High 'Snow Sonta', *Cosmos bipinnatus*

Flowering Tobacco, *Nicotiana balata*

Shasta Daisy, *Leucanthemum* x *superbum* 'Alaska'

Yes, I am a dreamer,
For a dreamer is one who can find his way by moonlight, and see the dawn before the rest of the world.
– OSCAR WILDE

Moonflower

Baby's Breath, *Gypsophila paniculata* 'Bristol Fairy'

Rose Campion, *Lychnis coronaria* 'Alba'

Four o'clocks, *Mirabilis jalapa*

Moonflower, *Ipomoea alba*

Monarda

Bee Balm, Bergamot, Fragrant Balsam, Oswego Tea, American Melissa, Square Stalk.

Monarda is a true native American herb, a member of the mint family and roughly 12 to 15 species can be found growing in North America. The two best known species in the wild and garden are *M. fistulosa*, commonly named wild bergamot, and the even more familiar *M. didyma*, or bee balm, a popular border plant introduced to gardeners in the seventeenth century.

Wild Bergamot

Both species display tubular whorls of terminal flowers on two to three foot distinctively square stems throughout July and August. The flowers of wild bergamot range from light lavender to whitish pink, while those of bee balm are brilliant red. Wild bergamot favors dry sunny localities from Maine to Minnesota and south to Florida and Kansas. The natural range of bee balm is more northerly than that of wild bergamot. It grows from Quebec and Michigan south to Tennessee and prefers moist organic soil.

Early explorers who came to America from England, France, Spain and Holland collected specimens of trees and plants to send home for study. The scarlet flowered bee balm, *M. didyma,* was one of the first American contributions to European herb and flower gardens. Spanish explorers sent the plant to Nicolás Monardes, an eminent physician and botanist living in Seville, Spain. Dr. Monardes probably called the plant bergamot because the leaves have a lovely sharp citrus and mint fragrance when crushed that is similar to bergamot oranges grown in Italy. The American Indians called this lovely native species of *Monarda O-gee-chee*, the "flaming flower."

American Indians used the leaves and flowers of *Monarda* for a variety of purposes, medicinally and as an insecticide. Dried *Monarda* leaves were mixed with pine needles and burned over coals for fragrance. The Oswego Indians of the eastern United States showed the colonial settlers how to make a minty tea from the plant's leaves. Oswego tea was used as a substitute for tea by rebellious Boston patriots during the 1773 China tea boycott, otherwise known as the Boston Tea Party.

Monarda can serve a multitude of uses in the home garden and naturalized settings. A large selection of colors and cultivars are available to meet the specialized needs of every gardener. The nectar of the flowers attracts bumblebees, butterflies, skippers and hummingbird moths. The Ruby-Throated Hummingbird also visits the flowers. White-tailed deer usually avoid them because of the minty-oregano flavor of the leaves.

Bee Balm is a wonderful plant for herb gardeners. The plants edible flower petals can be sprinkled on fruit salads and are very tasty added to apple jelly. Bergamot's citrus flavor blends well with pork dishes. The plant's fragrant leaves and flowers mix together with dried purple basil, calendula blossoms and rose-scented geraniums for a lovely natural potpourri blend.

GOOD FRIENDS — OUR NATIVE TREES

Sassafras *(Sassafras albidum)*

Young Sassafras Tree

Sassafras is a member of the native North American laurel family. It's a tall handsome deciduous tree that grows in open woods, rocky uplands and roadsides from Maine to Ontario and south to Florida and Texas. The sassafras tree is noted for it's aromatic wood, bark and roots and its long history of medical uses. Legend has it that the sassafras tree aided the

Europeans in the discovery of America. It was said that Christopher Columbus was assisted in his successful efforts subduing his homesick and mutinous crew because of the sudden fresh fragrance of the Sassafras tree which indicated that land was close at hand. The tree was highly prized by American Indians and was thought to cure an amazing range of aliments. The roots were boiled down into a strong tea to treat fevers and cleanse the blood. The bark and leaves were used to make medicinal teas and tonics. The leaves were dried for flavoring soups and the twigs were used as incense sticks. The flowers were mixed with beans before planting as a fertilizer.

During the colonial era, large amounts of the tree's aromatic root bark were shipped to Europe as a cure-all for everything from skin diseases to malaria. An oil extracted from its bark was used in dentistry, and as a flavoring for root beer and chewing gum until the early 1960s, when the United States Food and Drug Administration declared that safrole, a chemical compound found in the root was a potential carcinogen. Sassafras leaves are still used to make a seasoning called fil'e powder, an essential ingredient in Creole cooking.

Sassafras has bright greenish-yellow flowers in spring, brilliant red-orange leaves in the fall, green twigs in winter and an attractive horizontal branching habit. The sassafras tree enjoys the special distinction of bearing leaves of three different shapes on the same branch; oval, three-lobed and mitten-shaped. The tree produces beautiful dark, shining berries set on bright red stems that ripen in the fall. Birds and wildlife love the berries and quickly devour them as soon as they mature. The tree's flowers are a nectar source for the Spicebush and Pale Swallowtail butterflies.

Sassafras trees spread by underground runners and can be found growing in dense thickets in the wild. It is difficult to propagate and is rarely seen in nurseries. The *Brooklyn Botanic Garden All-Region Guide, Native Alternatives to Invasive Plants* recommends using this tree as a specimen, patio tree, screen or in naturalized settings.

▼▼▼▼▼▼▼▼▼▼▼▼▼▼▼▼▼▼▼▼▼▼▼▼▼▼▼▼▼▼▼▼

ACTIVITIES AND ADVENTURES FROM THE BACKYARD AND BEYOND

PURE FLOWER JUICE PAINTINGS

Flower Juice Painting

Materials Needed:

- Colorful fresh flower blossoms and fragrant leaves
- Absorbent paper
- A good imagination

Collect whole flower heads and herb leaves in the morning after the dew has dried. Salvia, marigold, calendula, larkspur, geraniums and dianthus flowers all produce bright and vivid colors. Fragrant leaves from scented geraniums, sage, lavender and mint plants create a lovely palette of green colors.

A porous absorbent paper works best for this project.

Use the flower blossoms the same way you would use small sponges or paintbrushes dipped in paint to add soft pastel colors to pencil drawings or create colorful, free-form paintings.

When gathering the flowers and leaves from the garden, discuss the different insects and birds that help pollinate plants and what purpose the color, shape and patterns of various flowers and leaves play in attracting pollinators to the flowers.

Save some of the different plants and press them for later. In the fall or winter make pressed flower pictures from the same flowers used for the summer flower paintings and talk about the different ways we use flowers for decorations and holiday celebrations.

Explore how plants got their names or the language of flowers. In Victorian days people would send small fragrant bouquets of flowers with hidden meanings called tussie-mussies. Show a few examples, such as marigold for happiness, rose for love, mint for cheerfulness and lavender for luck. Have them decode the secret messages

When the wind is in
the east, It's neither
good for man nor beast.
When the wind is
in the south,
The rain is in its mouth.
When the wind is
in the west,
It suits everyone best.

– TRADITIONAL LORE

of their paintings using Kate Greenway's *Language of Flowers*, or prepare a list from the Internet for them to use.

In a child's garden, imagination grows

– Gwen Frostic

Flower games help children learn about plant structure and form while having fun. The following game was published in *The Flower Grower*, a garden magazine from 1930. Children were assembled at a table and each child was given a sheet of paper and a pencil. A large pyramid of flowers, made of small bouquets, was placed in the center of the table. The children were asked to write a list of flowers, beginning with the letters of the alphabet in succession, such as Alyssum, Bluebell, Columbine, etc. A predetermined time limit was announced at the start of the game. The one who wrote the whole flower alphabet first got to select a bouquet to keep. As each player finished writing his or her list of flowers, they would win a chance to pick a bouquet. Those who were unable to finish their list had to pay a forfeit to win a bouquet; their forfeits were judged and redeemed before a bouquet could be taken. The game directions did not say what the forfeit was, so use your imagination depending on the children or group you are working with.

Flower Tower

Another fun game to play is called *Bouquet*. A leader is chosen, and then all the other players select a color, each one trying to choose a different color. The leader then says to the first player, "I am making a bouquet, how many flowers can you furnish?" The player then answers by naming all the flowers he or she can remember in the color he or she selected at the beginning of the game. A jellybean or some other type of counter is given for each flower named. At the end of the game whoever receives the most beans is the winner and becomes the new leader.

The flower painting and the related activities provide an interesting venue to introduce children to common garden and landscape flowers.

When you are looking for a fast fun summer project, close to home, that benefits the mind, body and spirit of children, this project is a hit with young and old alike!

THYMELY TIPS AND SAGE ADVICE

• Henry David Thoreau, July 12,1817 – May 6, 1862, is best known for his book *Walden*, which chronicles the two years that he lived in a small hut near Walden Pond in Concord, Massachusetts recording his observations of the region's native flora and fauna. Keep a family journal recording the evolving plants and animals in your backyard and beyond for future generations. Take pictures of the plants and wildlife at different times of the year. The American Museum of Natural History has a very informative Web page called How to Keep a Field Journal, http://www.amnh.org/nationalcenter/youngnaturalistawards/resources.

• *"Dog days of summer"* take their name from the rising of Sirius, the Dog Star in the constellation Canis Major, and is a term dating back to ancient Rome. Because the dog star rose and set with the sun during the six or eight hottest weeks of summer in July and August, they believed the star was responsible for the extra summer heat and humidity because its heat was combined with the sun's heat.

• Water and fertilize your window boxes and container gardens on a regular schedule. Encourage a longer blooming period for annuals and perennials by removing all faded flowers and foliage. When you water your lawn and garden let the water penetrate deeply into the soil rather than sprinkling lightly and frequently. Soil should be moistened to a depth of six inches.

• Garden pests are out in full force, but remember not all bugs are pests. By carefully monitoring your landscape and gardens from one year to the next and recording those observations, you will soon be able to predict and prepare for pest problems. Safe and effective pest control is a combination of cultural, biological, mechanical and chemical practices. Phenology is a useful tool in gardening to help predict insect problems. Insects are cold-blooded and respond to changes in weather conditions, and their growth and development is closely related to temperature. Plant-feeding insects have evolved over time with their host plants, so by closely monitoring those plants you will be able to control them when they first emerge.

AUGUST

In August, the large masses of berries,

which, when in flower, had attracted many wild bees,

gradually assumed their bright velvety crimson hue,

and by their weight again bent down

and broke their tender limbs.

– HENRY DAVID THOREAU

Sturgeon Moon — Red Moon — Green Corn Moon

Sturgeon is the common name of a large family of freshwater fish, of which various species were once found in great abundance along the Atlantic and Pacific coasts and parts of the interior of North America. The sturgeon is a primitive fish that has been around for thousands of years.

Instead of scales, sturgeons have five rows of bony plates call "scutes" and a long snout with slender whisker-like feelers that help them detect prey on murky bottoms of lake waters and rivers. They come in a variety of sizes and can reach lengths of over eight feet and weigh several hundred pounds. The American Indians named the full moon in August "Sturgeon Moon" because during this time the fish was plentiful in Northeastern regions of the United States.

Sturgeons were an important food source for American Indians and the dried fish was a highly valued product for trade. They harvested the fish using nets, spears, lassos and clubs. The fish were known for their curious nature and ability to leap up out of the water, occasionally landing right into a fisher's canoe. Sturgeons were found in great numbers when the colonists first arrived in the United States and legend has it there were so many sturgeon in the James River in Virginia you could cross the river by walking over their backs. The bountiful fish became a profitable export back to Europe. Colonists used the fish oil as a substitute for sperm whale oil because it was less smokey and had a longer storage life. A binding agent for paint was made from its bladder and the eggs of the sturgeon were used for caviar.

By the beginning of the nineteenth century there was a drastic drop in sturgeon populations due to overfishing, habitat destruction and blocked migration routes. Today the fish is protected, and environmental groups, federal agencies and individuals are working to restore their numbers and habitats. Families or school classes can help in these efforts by adopting a sturgeon and learning more about this interesting, prehistoric fish's role in American history, http://www.nero.noaa.gov/nero/education.

Some American Indian Tribes called August the time of the Full Red Moon. The moon tends to have a red-orange hue as it is rising and setting, which is caused by light passing through a higher concentration of dust particles in the earth's lower atmosphere. The saying *Red sky at night, sailor's delight, red sky in the morning, sailor take warning"* relates to pressure systems in the atmosphere which influence the concentration of dust particles at the horizon, or lower atmosphere. High pressure in the evening denotes fair weather coming from the west, thus the sky is red at night. If the sky in the eastern morning sky is red, the fair weather system has passed, with a low pressure system to follow.

The Full Green Corn Moon signaled that the first ears of corn would soon be ready for harvest. The Green Corn Ceremony or Dance was an annual celebration held by the Eastern Woodland and Southeastern tribes in late summer when the first corn ripened. No corn was eaten until the Great Spirit had been properly thanked. During the festival, members of the tribe would give thanks for the corn, rain, sun and a bountiful harvest. The dates and activities of the ceremony varied from tribe to tribe according to their traditions and when the first corn was ripe in their localities. Corn was an integral part of their daily lives and identities, "Every time we plant a seed, add compost to the soil, water a seedling, pull a weed, talk or sing gently to the plants or say 'thank you' for the blooming flowers, we are giving a gift. In turn we receive knowledge, peace of mind, food for our bodies, a growing spirit of giving and a sense of having a full life." (Caduto and Bruchac, 1996 p. 5) The American Indians have many legends concerning the origins of corn and they vary according to each tribe's traditions and geography. They all have a com-

'Red sky at night means fair weather tomorrow, red sky in the morning means foul weather all day.' You know how to interpret the weather signs in the sky, but you don't know how to interpret the signs of the times.
– Matthew XVI 2-3

mon theme of cultivating a sense of gratitude and respect for what they had been blessed with in their lives and the importance of passing these values to their children.

August is a time of transition in the circle of the seasons. If you take time to read the landscape in your backyard and beyond you will see the subtle changes taking place as Mother Nature moves from summer to autumn. Wildflowers reach their peak by late summer and the rhythmic sounds of insects in the hedgerows, meadows and trees tops have replaced the songs of birds.

Wildflowers appear in orderly succession of bloom from spring through fall, beginning with the simple, softly-colored flowers and finishing with blazing yellow, golds, vivid blue and scarlet-colored shades. Many of the plants we see blooming in late summer and early fall are all members of the composite family, the largest and most highly developed of all plant families. The composites are a diverse family of plants with varied and interesting flowers that are able to adapt to a wide range of locations and soil types. Asters, goldenrods, begger-ticks and chicory all belong to the composite family.

No matter what time of day you go outside in August, you can see or hear insects munching, humming, clicking, sucking, weaving webs, stitching or mining leaves and pollinating flowers. Everyone is storing food and preparing comfortable lodging for their families to spend the cold winter days.

How to Spend a Lazy Summer Day

Morning

Look for: Freshly woven spider webs spangled with drops of dew on lawns, shrubs and plants. Spiders that hatched back in the first warm days of spring are now large enough to be easily observed spinning their

> When summer opens, I see how fast it matures, and fear it will be short; but after the heats of July and August, I am reconciled, like one who has had his swing, to the cool of autumn.
>
> – RALPH WALDO EMERSON

webs, hunting, eating insects and laying eggs. Spiders are not insects, they are arachnids. Insects have six legs and three main body parts — a head, thorax and abdomen. Spiders have two main body parts, a head and abdomen with eight legs. Spiders do not have antennae or wings, but they do have eight eyes, fangs and special glands in their abdomen that make silk. If you look at a spider's eyes with a hand lens you will see that they aligned in interesting patterns.

Spiders weave several styles of webs easily identified early in the day before the dew dries on them. Sheet webs look like shallow silk bowls attached to blades of grass or branches of trees and shrubs. Funnel or grass spiders construct large, flat horizontal webs with holes in their centers. The orb weavers build round webs and hang them vertically between tall plants in vacant fields and wild patches along roadsides.

Spiders can be divided into one of two groups depending how they catch their prey. The hunting or wandering spiders rely on their speed and vision to capture prey and the web builders patiently wait for food to come to them. Not all spiders make webs, but they are all able to make silk, and use it to wrap prey, line burrows and create egg sacs. Spider-WebWatch.org. is an interactive Citizen Science program about spiders with more in-depth information and fun family activities to do with spiders.

Some folks believed that if one kills a spider it won't rain for seven days. "If you want to live and thrive, Let the spider run alive." Children living on farms in times past were careful not to harm a spider in dry weather. **Look for:** During the day which flowers open first in the day and which ones open later? Compare the opening of the flowers with different times friends or family members get up in the morning.

Listen for: What are the first sounds you hear in the morning? How many birds can you hear? How have their songs changed since they ar-

Garden Spider

CHIEF SEATTLE
All things are
connected.
What ever befalls the
earth befalls the sons of
earth. Man did not
weave the web of life;
he is merely a strand
in it. What ever he
does to the web,
he does to himself.

rived in spring?

Try this: Find a smooth, solid surface outside and trace your child's shadow with a large piece of chalk. Then switch places and have them trace your shadow. Add some details to the shadows. Next, try tracing the shadows of trees and shrubs along the sidewalks and parking lots in your yard and neighborhood. Add some details to them such as bugs, berries and birds. Are shadows longer in summer or winter?

Afternoon

Look For:

Three different types of leaf edges

A rock with a pattern on it

A living fossil

A whispering sound

A waxy surface

A reflection

Something that prickles

Five shades of green

Something older than you

Sun on a leaf

A spongy piece of wood

Three colors of soil

Something that makes you happy

Something that looks slimy

Something that comes in pairs

Something shiny

A home that is hidden

A sweet smell

A colorful leaf

Something that swims

Something you can't count

Look for: Evidence of insects

Investigate the green leaves of the shrubs, trees and plants in your yard and neighborhood for evidence of insect activity. Just as people have favorite foods, so do insects. They vary in their eating habits, but each insect type has its own distinctive style. Entomologists are able to identify insects by studying how they use plant material. Small bites missing out of leaves are usually evidence of moth and butterfly caterpillars. The twisting lines and blotches on leaf surfaces indicate leaf miners at work. Experts are able to identify these insects by the form of their 'signature' pattern on leaves. Some insects roll, tie or fold leaves for shelter and hibernation. If you want to see how an insect sews or rolls a leaf, carefully unfold or untie one and place the insect on a new leaf. Hydrangeas, a common variety of shrubs in home landscapes, occasionally host a little green caterpillar that likes to sew leaves together.

Galls, strange bumps and spheres on leaves and stems of plants are caused by insect eggs. Some moths, beetles, tiny aphids, flies and wasps make galls in all sizes, shapes and colors. Wilted leaves or vines are most often caused by leaf and stem borers, the larva of moths or beetles. You might have noticed that some leaves do not show any evidence of insect damage. What characteristics do these leaves have that might make them unattractive to insects?

Listen for: The Insect Orchestra

Birds, mammals, toads and frogs are not very active on hot sunny afternoons. Apart from insects, snakes are about the only wildlife out and about during the hazy, hot days of August. Make a list of the different types of sounds you can hear and match them with the proper owner.

SUMMER SOUNDS

Chirping

Clicking

Thunder

Falling rain

Tree frogs

PATTERNS

Sun beams

Composite flowers

Ripe seed heads

Shades of green

Shadows

Spider webs

Grasshoppers, cicadas, katydids and click beetles live in fields, under leaves and rocks and the tall grass in your backyard and beyond. Listen carefully to their songs and let the sounds direct you to their hiding spots. Each one makes a distinctive rhythmic sound by rubbing their legs or wings together. When blended they sound like musicians playing an ode to fall.

According to old weather lore, after you hear the annual cicadas sing for the first time, a killing frost will come in about six weeks. This bit of weather lore does have some merit, since annual cicadas usually emerge from the soil each summer during the dog days of summer. This is only a few months from the first frost in Midwestern and Northeastern regions of the United States. Jarfly, dog-day, harvest fly and locust cicada are a few other common names of the cicada. Mark your calendar when you hear them for the first time and begin the countdown to the first frost.

Night

Listen for: The Music of the Night

Think about the sounds of spring and early summer and compare the sounds of an August night with a night in early May. The trilling green tree cricket, a rarely seen insect, sings one of the most common insect songs people hear after dusk in August. The gray tree frog's loud resonant trill is often heard in accompaniment with the tree cricket. If you go out at night with a flashlight and listen carefully you can sometimes track them by their songs. The gray tree frog likes to sit on the house roofs at night and the green cricket perches in leafy green shrubs and trees. To hear the voice of the gray tree frog visit nwf.org/frogwatchUSA.

The black field cricket makes a cheerful chirping sound and lives under rocks and plant debris on the ground. It has been said that by listening to the frequency of a black cricket's chirps you will be able to get a rough estimate of the temperature. Count the number of chirps in 15 seconds

And there's never a leaf nor blade too mean, to be some happy creature's palace.

– JAMES RUSSELL LOWELL

Green Tree Cricket

Tree Frog

and add 37 and the number you get will be an approximation of the outside temperature.

Jewelweed *(Impatiens capensis)*

Jewelweed is a fascinating and fun plant that is best observed where it grows. Its buds, flowers, leaves, seed capsules and root system all have distinctive and noteworthy features. Silver cap, touch-me -not, lady's eardrops and snapweed are common names attached to this magical plant. Each name provides a clue to an interesting aspect of the plant.

The most common species of jewelweed found growing in the Eastern regions of the United States is spotted touch-me-not, *I. capensis*, a leafy annual that grows in wet, shady places. It has orange-yellow flowers spotted with reddish-brown that bloom in midsummer and hang like delicate jewels on a lady's necklace. The trumpet-shaped flower has three sepals, one which curls around to form a nectar sac with a long spur, especially adapted for a hummingbird's beak. A variety of insects can be seen flying about the bright orange flowers or munching on the leaves on warm sunny days in late summer. If you look closely you will notice "nectar thieves", primarily bees, stealing around to the back of the jewelweed's flower blossom and shamelessly siphoning nectar directly from its storage sac.

Jewelweed is a tall fragile plant, with succulent stems and a shallow-spreading root system. Plants grow together in dense patches with their roots overlapping helping to support one another. Its delicate oval, scalloped leaves shed water, which collects

Jewelweed or Snapweed

MAGIC JEWELS

She brooks no
condescension from
mortal hand, you know,
For, touch her
e'er so gently,
impatiently she'll throw
Her tiny little jewels,
concealed in
pockets small
Of her dainty, graceful
garment, and o'er the
ground they fall.
Her tiny magic jewels
may be a fairy's gift,
For scattered by the
brook side they soon
small leaflets lift.
What mortal knows the
secrets of Flora's
children shy,
Concealed in field and
meadow, that with
the flowers die?
— RAY LAURANCE

along the top side of the leave's notched edges like crystal beads. If you submerge the plant's dark green leaves under water, the undersides take on the color of gleaming silver, hence the name silver-cap.

When disturbed, the ripe seed capsules explode, propelling beautiful blue seed as far as four or five feet from the mother plant. Children like to play games with the seed pods, taking turns touching the mature pods to see who could explode and scatter the seeds the farthest distance. The ripe seeds can be used as a substitute for walnuts in cookies and cakes. Collect the exploding seeds by tightly securing a paper bag over the flowers of several plants and shake them to release seeds into the bag.

American Indians used juice from this plant as a remedy for skin rashes, poison ivy, insect bites and as a fungicide. Many people still use jewelweed as antidote for poison ivy.

The following recipe is used with permission from the *Creative Herbal Home* by Susan Belsinger and Tina Marie Wilcox

Jewelweed Vinegar

1 cup fresh crushed jewelweed
2 cups apple cider vinegar

Place jewelweed in glass quart jar. Cover with vinegar and seal with a plastic lid (vinegar corrodes metal). You can use it in a day or two or leave the herb in for up to four weeks. Pour vinegar through a cheesecloth-lined strainer. We add insect repellent and antiseptic essential oils to the vinegar, ten drops to each one-pint sprayer. The spray is kept nearby to subdue itchy fits and to reapply insect repellent oils. As a variation, we make Herbal Insect Repellent Vinegar to mix with the Jewelweed Vinegar. The vinegars are good for about one year.

ACTIVITIES AND ADVENTURES FROM
THE BACKYARD AND BEYOND

PLANT WEAVINGS

Tree branches, dry weeds, grasses, leaves, vines and roots can all be used to create beautiful nature weavings on natural or manmade looms. Weaving is enjoyable — the more you practice the better you become!

BRANCH WEAVING

Materials needed:

- Yarn or string
- Forked tree branch
- Grasses, leaves, flower tops and twigs

Branch Weaving

Tie one end of the yarn to one forked branch close to the bottom where the branches join and then wrap it once around the branch. Stretch the yarn back across to the other branch and wrap it around once. Go back to the first branch and wrap the yarn about a quarter of an inch below the first wrap. Continue taking the yarn back and forth, working toward the wide part of the forked branch. Tie the yarn to one of the branches a couple of inches from the top and cut off the excess yarn. Weave the plant materials up and down going over one strand of yarn and under the next.

HERBAL WEAVING

Materials needed:

- Burlap fabric one-foot wide by one-foot long
- 16-inch long and 1/2-inch diameter wide branch
- Grasses, weeds and fragrant herbs with long stalks
- String
- Two dried poppy heads

Herbal Weaving

Take a piece of burlap fabric, at least one-foot wide and one-foot long. Fold over the fabric top one inch and glue or sew across the length of the fabric to create a holder for the stick.

Pull out eight or more of the horizontal burlap strings in five different places in the fabric to create a loom for weaving in the plant material. Gather grasses, weeds and herbs with long stalks to use for weaving.

Weave the plants over and under through the spaces where the burlap strings have been removed. Try to be neat on both sides of the material. When the project is finished there will be a right side. The fragrant herbs are fun to work with and add a delightful lasting fragrance to project.

Insert the stick through the top of the finished weaving and glue the dried poppy heads to both ends of the stick to create ornamental finials. Tie on a string and hang on a wall.

These two simple projects are a great way to encourage children down a path of self discovery by progressively increasing their weaving skills with more difficult projects and techniques. Cornell University has an interactive Web site called *Garden-Based Learning* at http://blogs.cornell.edu/garden. The site has information using plants for dyeing, mat weaving and rope making and has good tips on how to teach these skills to younger children.

Basket Weaving

The primitive little basket pictured here was woven by Sherry Tucker from flexible vines collected in her yard in Delaware. An avid gardener and basketmaker, she creates whimsical baskets from a variety of natural materials collected from her backyard and beyond. For step by step instructions to make this advanced basket project visit *Practical Primitive*, a Web site dedicated to teaching valuable traditional and primitive arts, crafts and outdoor skills. It is a very good resource for educators working with older children and adults, http://www.practicalprimitive.com/melonbasket.html.

GOOD FRIENDS — OUR NATIVE TREES

Tulip Tree *(Liriodendron tulipifera)*

The tulip tree is one of the most impressive and magnificent native trees inhabiting the forests and woodlands of Eastern North America. *Liriodendron* is a genus of two species of trees belonging to the Magnolia family, one in the United States and the other in central China.

It is also known as whitewood, yellow poplar, and tulip poplar, which describes the fluttering of its leaves like those of the poplar. Henry Ward Beecher once said, *"Under the same wind one is trilling up and down, another is whirling, another vibrating right and left, still others are quieting themselves to sleep."*

The name whitewood refers to the color of the trunk's straight-grained inner wood, highly valued by the American Indians to craft their dugout canoes. They made an infusion from the root bark to treat snake bites and the flowers were valued for honey production.

Tulip Tree Flower

Few people are familiar with the tree's lovely greenish-white tulip-shape flowers, which bloom high on the tree after the leaves emerge in late May and early June; hence its most common name, tulip tree. One of the tallest of our native broad-leaved trees, records dating back to the arrival of the first settlers describe tulip trees 200 feet tall and more than twelve feet in diameter. Botanists sent specimens of the tree back to Europe and today it's one of the more familiar American trees in European landscapes. Tulip poplars were used as an indicator species by early colonists because they noticed them growing in deep, rich well-drained soils perfect for farming. Lumbermen called it yellow poplar because they found its yellow heartwood to be very versatile and used the wood for making furniture, boat building, shingles, weather board and wooden ware. The yellow poplar or tulip poplar is still valued today in the lumber trade but has become high priced and scarce.

The tulip tree is a fast-growing shade or ornamental tree and can easily reach fifteen to twenty feet in six to eight years. The large size makes it unsuitable for small spaces. Its fine points are the tree's stately columnar trunk, pyramidal canopy of unusual-shaped leaves, attractive flowers and superb yellow fall color. A variety of wildlife like to munch on its twigs, fruit and seeds. Honeybees love the flowers in spring and produce a dark honey that is especially favored by bakers. The tulip tree is the host plant for the Tiger Swallowtail butterfly.

"The tulip tree is always the well balanced, upstanding American, a leader in any company," said Robert Lemmon, a famous writer for gardening magazines in the 40s. The historical value and majestic character of the tulip tree entitle it to a place of honor in America's community parks, schoolyards and neighborhood green spaces.

THYMELY TIPS AND SAGE ADVICE

• Watering is a primary task during the hot dry days of August. Be sure to water thoroughly and deeply each time. This will encourage plants to develop deeper root systems that will withstand hot dry spells. Surface watering wastes water because it doesn't reach the plant's root zone and the moisture quickly evaporates from the top inch of soil.

• Divide and transplant spring and early summer blooming perennials. This is least stressful on the plants when done on a cool, cloudy day. Be sure to keep plants well watered after transplanting.

• Late summer is a good time to find bargains on container-grown trees and shrubs at your local nursery. Be sure to take time to properly prepare the soil before planting by mixing generous amounts of compost with your existing soil. Soak the plant thoroughly before removing it from the container and again after planting. Check to make sure the roots of the plant are not tightly bound; gently loosen them apart before placing them in the ground. Keep plants regularly watered to help them get off to a healthy start.

• Water, fertilize and deadhead your annual flowers to keep them looking great until frost. Remove spent blooms and prune back plants that start to look tired. Use a water soluble fertilizer every two weeks to keep plants blooming. Leave some seed heads to self sow next spring.

• Fish Bait — For catching fish in large quantities, take equal parts of the following herb seeds: lovage, fennel, cumin, coriander and anise and mix throughly. Steep 7 teaspoons of this mixture in a cup of water on the back of the stove for one hour; then strain. When cold put a few drops on any bait. From *The Herb Doctor and Medicine Man*, by Joseph Meyer, 1922

• A tough winter is ahead if corn husks are thick and tight; birds migrate early; squirrels tails are very bushy; berries and nuts are plentiful; and bees build their nest high in trees.

Autumn

The morns are meeker than they were,
The nuts are getting brown;
The berry's cheek is plumper,
The rose is out of town.

The maple wears a gayer scarf,
The field a scarlet gown.
Lest I should be old fashioned,
I'll put a trinket on

– Emily Dickinson

Shin-Ge-Bis Fools The North Wind

Long, long ago in the time when only a few people lived upon the earth, there dwelt in the North a tribe of fishermen. Now, the best fish were to be found in the summer season, far up in the frozen places where no one could live in the winter at all, for the King of this Land of Ice was a fierce old man called Ka-bib-on-okka by the Indians — meaning, in our language, "the North Wind."

Though the Land of Ice stretched across the top of the world for thousands and thousands of miles, Ka-bib-on-okka was not satisfied. If he could have had his way there would have been no grass or green trees anywhere; all the world would have been white from one year's end to another, all the rivers frozen tight, and all the country covered with snow and ice.

Luckily, there was a limit to his power. Strong and fierce as he was, he was no match at all for Sha-won-dasee, the South Wind, whose home was in the pleasant Land of the Sunflower. Where Sha-won-dasee dwelt it was always summer. When he breathed upon the land, violets appeared in the woods, the wild rose bloomed on the yellow prairie, and the cooing dove called musically to its mate. It was he who caused the melons to grow and the purple grapes; it was he whose warm breath ripened the corn in the fields, clothed the forests in green, and made the earth all glad and beautiful. Then, as the summer days grew shorter in the North, Sha-won-dasee would climb to the top of a hill, fill his great pipe, and sit there dreaming and smoking. Hour after hour he sat and smoked; and the smoke, rising in the form of a vapor, filled the air with a soft haze until the hills and lakes seemed like the hills and lakes of a dreamland — not a breath of wind, not a cloud in the sky —a great peace and stillness over all. Nowhere else in the world was there anything so wonderful. It was Indian Summer.

Now it was that fishermen who set their nets in the North worked hard and fast, knowing the time was at hand when the South Wind would fall asleep, and fierce old Ka-bib-on-okka would swoop down upon them and drive them away. Sure enough, one morning a thin film of ice covered the water where they set their nets! A heavy frost sparkled in the sun on the bark roof of their huts.

That was sufficient warning. The ice grew thicker; the snow fell in big, feathery flakes. Coyote, the prairie wolf, trotted by in his shaggy, white winter coat. Already, they could hear a muttering and a moaning in the distance.

"Ka-bib-on-okka is coming!" cried the fishermen. "Ka-bib-on-okka will soon be here. It is time for us to go."

But Shin-ge-bis, the diver, only laughed.

Shin-ge-bis was always laughing. He laughed when he caught a big fish, and he laughed when he caught none at all. Nothing could dampen his spirits.

"The fishing is still good," he said to his comrades. "I can cut a hole in the ice and fish with a line instead of a net. What do I care for old Ka-bib-on-okka?"

They looked at him with amazement. It was true that Shin-ge-bis had certain magic powers and could change himself into a duck. They had seen him do it, and that was why he came to be called the "diver"; but how would this enable him to brave the anger of the terrible North Wind?

"You had better come with us," they said. "Ka-bib-on-okka is much stronger than you. The biggest trees of the forest bend before his wrath. The swiftest river that runs freezes at his touch. Unless you can turn yourself into a bear or a fish, you will have no chance at all."

But Shin-ge-bis only laughed the louder.

"My fur coat lent me by Brother Beaver, and my mittens borrowed from

Cousin Muskrat, will protect me in the daytime," he said, "and inside my wigwam is a pile of big logs. Let Ka-bib-on-okka come in by my fire if he dares."

So the fishermen took their leave rather sadly, for the laughing Shin-ge-bis was a favorite with them; and the truth is, they never expected to see him again.

When they were gone, Shin-ge-bis set about his work in his own way. First of all he made sure that he had plenty of dry bark and twigs and pine needles to make the fire blaze up when he returned to his wigwam in the evening. The snow by this time was pretty deep, but it froze so hard on top that the sun did not melt it, and he could walk on the surface without sinking in at all. As for fish, he well knew how to catch them through the holes he made in the ice; and at night he would go tramping home, trailing a long string of them behind him and singing a song he had made up himself:

"Ka-bib-on-okka, ancient man,
Come and scare me if you can.
Big and blustery though you be,
You are mortal just like me!"

It was thus that Ka-bib-on-okka found him, plodding along late one afternoon across the snow.

"Whoo, whoo!" cried the North Wind. "What impudent, two-legged creature is this who dares to linger here long after the wild goose and the heron have winged their way to the South? We shall see who is master in the Land of Ice. This very night I will force my way into his wigwam, put his fire out, and scatter the ashes all around. Whoo, whoo!"

Night came; Shin-ge-bis sat in his wigwam by the blazing fire, and such a fire! Each backlog was so big it would last for a moon. That was the way the Indians, who had no clocks or watches, counted time: instead of weeks or months, they would say "a moon," the length of time from one new moon to another.

Shin-ge-bis had been cooking a fish, a fine fresh fish caught that very day. Broiled over the coals, it was a tender and savory dish; and Shin-ge-bis smacked his lips and rubbed his hands with pleasure. He had tramped many miles that day, so it was a pleasant thing to sit there by the roaring fire and toast his shins. How foolish, he thought, his comrades had been to leave a place where fish were so plentiful so early in the winter.

"They think that Ka-bib-on-okka is a kind of magician," he was saying to himself, "and that no one can resist him. It's my own opinion that he's a man just like myself. It's true that I can't stand the cold as he does; but then, neither can he stand the heat as I do."

This thought amused him so that he began to laugh and sing:

> *"Ka-bib-on-okka, frosty man,*
> *Come to freeze me if you can.*
> *Though you blow until you tire,*
> *I am safe beside my fire!"*

He was in such a high good humor that he scarcely noticed a sudden uproar that began without. The snow came thick and fast. As it fell, it was caught up again like so much powder and blown against the wigwam, where it lay in huge drifts; but instead of making it colder inside, it was really like a thick blanket that kept the air out.

Ka-bib-on-okka soon discovered his mistake, and it made him furious. Down the smoke vent he shouted, and his so wild and terrible that it might have frightened an ordinary man, but Shin-ge-bis only laughed. It was so quiet in that great silent country that he rather enjoyed a little noise.

"Ho-ho!" he shouted back. "How are you, Ka-bib-on-okka? If you are not careful, yo will burst your cheeks."

Then the wigwam shook with the force of the blast, and the curtain of buffalo hide that formed the doorway flapped and rattled and rattled and flapped.

"Come on in, Ka-bib-on-okka!" called Shin-ge-bis merrily. "Come on in and warm yourself. It must be bitter cold outside."

At these jeering words, Ka-bib-on-okka hurled himself against the curtain, breaking one of the buckskin thongs, and made his way inside. Oh, what an icy breath! — so icy that it filled the hot wigwam like a fog.

Shin-ge-bis pretended not to notice. Still singing, he rose to his feet and threw on another log. It was a fat log of pine, and it burned so hard and gave out so much heat that he had to sit a little distance away. From the corner of his eye, he watched Ka-bib-on-okka, and what he saw made him laugh again. The perspiration was pouring from his forehead; the snow and icicles in his flowing hair quickly disappeared. Just as a snowman made by children melts in the warm sun of March, so the the fierce old North Wind began to thaw! There could be no doubt of it: Ka-bib-on-okka, the terrible, was melting! His nose and ears became smaller and his body began to shrink. If he remained where he was much longer, the King of the Land of Ice would be nothing better than a puddle.

"Come on up to the fire," said Shin-ge-bis cruelly. "You must be chilled to the bone. Come up closer and warm your hands and feet."

But the North Wind had fled even faster than he came through the doorway.

Once outside, the cold air revived him, and all his anger returned. As he had not been able to freeze Shin-ge-bis, he spent his rage on everything in his path. Under his tread, the snow took on a crust; the brittle branches of the trees snapped as he blew and snorted; the prowling fox hurried to its hole; and the wandering coyote sought the first shelter at hand.

Once more, he made his way to the wigwam of Shin-ge-bis and shouted down the flue. "Come out," he called. "Come out if you dare, and wrestle with me here in the snow. We'll soon see who's master then!"

Shin-ge-bis thought it over. "The fire must have weakened him," he said to himself, "and my body is warm. I believe I can overpower him. Then he will not annoy me anymore, and I can stay here as long as I please."

Out of the wigwam he rushed, and Ka-bib-on-okka came to meet him. Then a great struggle took place. Over and over on the hard snow they rolled, locked in one another's arms.

All night long they wrestled; and the foxes crept out of their holes, sitting at a safe distance in a circle, watching the wrestlers. The effort he put forth kept the blood warm in the body of Shin-ge-bis. He could feel the North Wind growing weaker and weaker; his icy breath was no longer a blast but only a feeble sigh.

At last, as the sun rose in the East, the wrestlers stood apart, panting. Ka-bib-on-okka was conquered. With a despairing wail, he turned and sped away. Far, far to the North he sped, even to the land of the White Rabbit; and as he went, the laughter of Shin-ge-bis rang out and followed him. Cheerfulness and courage can overcome even the North Wind.

Larned, W.T.L., *American Indian Fairy Tales*, P.F. Volland Company,1921. (With one exception, all the tales in this book are adapted from the legends collected by Henry J. Schoolcraft.)

SEPTEMBER

Now in the ninth month the great clan of Compositae
comes to glory in a burst of gold and purple.
Of all the families it is not only the largest, and the newest
in point of evolution, but at the very peak and summit
of the plant kingdom…so just as the year opens
with the most primitive and ancient families,
like the catkin-bearing trees, it ends with a triumph
of the newest and most complex flower families.

– D.C. Peattie

Harvest Moon — Grasshopper Moon

Summer ends and autumn begins at the autumnal equinox, sometime between September 21 and 23 each year in the Northern Hemisphere. The full moon nearest the autumn equinox is called the Harvest Moon, which can occur in September or October. Typically the full moon rises an average 50 minutes later each night, but for the few nights near the time of the autumn equinox, the moon rises only 30 minutes later each day.

*By all the lovely tokens
September days are here,
With summers
best weather
And autumns best cheer.*

– HELEN HUNT JACKSON

It's more than just a connection to the season of harvest. In fact, nature is particularly cooperative during the months of autumn to make the Harvest Moon unique. The shorter-than-usual time between the moonrises around Harvest Moon time means there is no long period of darkness between sunset and moonrise. In the days before tractor light, the light of the autumn full moon helped farmers bring in their crops. As the sun's light faded in the west, the full moon would rise in the east to illuminate the fields for several more hours of work. (Deborah Byrd, September 22, 2010) *EarthSky: A Clear Voice for Science*, www.Earthsky.org, is an informative Web site that explains astronomy topics in simple terms.

American Indians planted corn, beans, squash and sunflowers, which were ready for harvest during this moon month. A good portion of these crops were dried and stored in caches for winter food. Wild vegetables readily available without cultivation were also gathered and dried for future use, such as Broad-Leaved Arrowhead or "wild potato" and Jerusalem artichoke roots. Moss growing on white pine trees was collected and stored to freshen water during wintertime. Wild rice was gathered in the fall, and just like maple sugar time, it was an essential seasonal task as well as a pleasant social time in the cycle of the seasons.

American Indians gathered plants throughout the spring, summer and fall. They had an intimate knowledge of their local environment and paid careful attention to how the day and the seasons affected the plants they gathered. They knew when each plant was at its peak of potency or perfectly ripe. Special attention was paid to this activity in late summer when many plants were fully developed and in blossom. Roots were gathered in early spring or late fall and bark in summer "when the sap was in the tree." Tobacco was always placed on the ground before roots, flowers or bark were harvested, the tobacco being first offered to the cardinal points, the zenith and earth. They would speak in a respectful voice saying the plant material was being taken for a useful purpose. They would be careful to only gather what was necessary and ask that its use be successful. (Densmore, p.127)

Tobacco Bundle

A great many of the insect sounds we hear in late summer and early fall are made by grasshoppers and crickets. Grasshoppers are around in the spring and summer, but are more noticeable in the autumn. Grasshoppers are divided into two groups: short- and long-horned families in the order *Orthoptera*. The horns or antennae are considered long if they are nearly as long as or longer than the insect's body. The two groups of insects produce different sounds. It has been said the short-horned grasshoppers shuffle, rustle or crackle and long-horned grasshoppers scratch and scrape. If you listen carefully you can learn to recognize their songs just as you can identify birds by their songs. The sounds are usually made by males trying to attract a mate.

The poetry of the earth
is never dead;
When all the birds
are faint with sun
And hide in cooling trees,
a voice will run.
From hedge to hedge
about new-morn mead
That is the
grasshoppers.
– KEATS

Grasshoppers have three stages of growth. Eggs are laid in late summer and winter over and hatch the following spring. After hatching, the young insect that emerges is called a nymph. A nymph looks just like a tiny adult, without wings. The nymph reaches adulthood when it grows wings and can fly. Birds, black bears, lizards and mantids, spiders and rodents eat grasshoppers. The Lenni Lenape called this moon Grasshoppers Moon because these insects were easy to see and hear in late summer and early fall singing, hopping and crawling in the grassy fields.

Herbs

Whether used in medicine, foods, dyes, decorations, cosmetics or cultural traditions, herbs have played an integral role in the lives of mankind before the earliest recorded history. Herbs are useful plants imbued with centuries of folklore and superstition. Many of these beliefs center on their medicinal qualities and others on the magical properties attributed to them. With the current advances in science, some, but not all of these powers have been disproved. There is still much to be discovered about the wonderful abilities of herb plants. Herbs are history plants and each one has a story to tell.

The original American herb gardens were the native plant communities naturally existing in the fields, woods, meadows and wetlands across North America. Early settlers learned much about the native herbs growing in America by watching and listening to the American Indians.

Colonists brought the seeds and roots of their favorite herbs, flowers, grains and vegetables to the New World. They found great comfort and companionship in their herb and flower gardens. Growing and tending the familiar plants from their homelands kept them connected to their cultural traditions. Their plants adapted well to their new homes and many of these green immigrants have become naturalized members of native plant communities across America.

One of the plants the early settlers brought to America was a perennial herb called plantain *(Plantago major)*. The American Indians observed that wherever the white man went this plant would soon appear, giving it the common name White-man's foot. They quickly recognized plantain's edible and medicinal value. The leaves were used in spring salads or boiled as a potherb. It was a highly respected herb in its homeland long ago. Today it is considered by some to be a noxious weed while others have great faith in its healing qualities.

In Marche and in April,
from morning to night;
in sowing and setting,
good huswives delight.

To have in their gardein
or some other plot:
to trim up their house,
and to furnish the pot.

At Spring (for the somer)
sowe garden ye shall,
at harvest (for winter) or
sowe not at all.

Oft digging, removing
and weeding (ye see)
makes herbe the more
wholesome and
greater to be.

– Thomas Tusser
Five Hundred Points of
Good Husbandry, 1573

Michigan herbalist Coleen French created the following Plantain Salve recipe, a simple remedy to treat skin ailments. She discusses this recipe below.

PLANTAIN HERB SALVE RECIPE

Plantain *(Plantago* spp*.)* is an herb that grows freely in the fields and meadows around my home. I have found it wherever I have travelled, including along public streets, even growing vigorously in cracks in the sidewalks. Plantain has been used for centuries to heal wounds, itch rashes, minor burns, and insect bites. The common or the narrow-leaf plantain will work equally well. This is a very good salve to have in your herbal first aid kit. A salve is less messy than an oil, and far more portable.

A salve is made much like a tea and extra virgin olive oil is used instead of water to extract the properties of the herbs (infusion). After the infusion process is complete, the herbs are strained out of the oil. The oil is then warmed with beeswax to thicken the oil to salve consistency.

Pack the herbs into a glass canning jar, then cover the herbs completely with oil. Leave as little air space at the top as possible, seal, and place in a sunny windowsill for two to three weeks, shaking daily.

Plantain Plant
Photo courtesy
of Coleen French

If you are using fresh herbs, let them wilt for a day or two to remove excess water. This helps lessen possibility of mold forming. If using dried herbs, fill the jar only about half full of herbs, then add oil to cover the dried herbs. Let sit for about an hour. The dried herbs will absorb the oil and expand. Now you can see how much more oil to add to fill the jar, or if you need to add more dried herbs. If you fill the jar to the top with dried herbs, and then add oil to fill, you will end up with herbs expanding right out of the jar!

In two to three weeks, strain the herbs out of the oil. Cheesecloth or other fine fabric, like that used for 'sheer' window coverings, works well for this. Add beeswax to the oil, in a proportion of about one ounce of beeswax to 1 cup of infused oil, depending on how stiff or soft you want your salve. Adding some Vitamin E will help preserve your salve. Pour the melted salve/beeswax mixture into small clean jars, cool and label carefully with the herb name and date prepared, and brief instructions for use.

Harvesting, Drying and Preserving Herbs

Depending on the plant you have grown or wish to gather from the wild, the time of season and the plant's life cycle will help guide you in determining what part to harvest and preserve for later use.

Herbs grown for the flavor of their leaves are generally strongest just before the plants flower. Harvesting can begin any time there is sufficient growth on the plants so they can withstand cutting. Never cut a plant completely back except in the case of annuals at the end of the season. Harvest herbs in the morning after the dew has dried. Some domestic herbs grown by the colonists for their flavorful leaves included parsley, sage, savory, rosemary, marjoram, lavender, thyme and lemon verbena.

Flowers can be harvested just before they are fully open. Cut the flower heads off and thinly spread them on screens or racks. A few herbs grown for their flowers include borage, chamomile, Dianthus, roses and violets.

Seeds should be harvested when they are fully ripened. You can cut the whole plant or just the flower stalk. Hang the plants upside down to dry enclosed in paper bags to catch the seeds. Some herbs grown for their seeds include dill, fennel, coriander, anise, sunflower, cumin and caraway.

Harvest root crops in the fall when the plant parts are beginning to wither and dry. Gather perennial herbs in their second or third year. Carefully dig the whole root and separate the amount needed and replant the remainder of the root. Throughly wash and scrub the roots to remove the dirt. Split or slice the large roots and spread in thin layers on drying screens in the open air. The colonists harvested the roots of lovage, angelica, comfrey, horseradish and sweet flag for a multitude of uses.

Large-leafed herbs can be hung in loose bunches in a dark dry warm place. It can take from two days to several weeks for the herbs to completely dry.

Store dried herbs in clean glass jars away from heat and light to preserve their flavor, fragrance and color.

Goldenrod *(Solidago* sp.*)*

The goldenrods *(Solidago* sp.*)* are members of a genus of plants in the family Asteraceae, native to North America. They are commonly found as wildflowers in their native range, but some are also grown as garden plants. Depending on which source you read, there are 60 to 90 varieties, many of them quite similar and hard to tell apart. Goldenrods bloom in clusters of tiny yellow-rayed flowers on long stalks. Peterson's *A Field Guide to Wildflowers* recommends identifying the plants by their flower shapes. There are those with plum-like graceful clusters, elm-branched plumes, club-like, showy clusters topping stems, small wand-like clusters and flat-topped clusters shaped like a bright yellow Queen Anne's Lace.

Goldenrod does not cause seasonal allergies. It has heavy, sticky pollen that is carried from flower to flower by insects, not the wind. The dully-colored, wind-pollinated ragweed, growing in the same season as goldenrod, gives goldenrod its ill-founded unpleasant reputation.

American Indians introduced the colonists to the plant and its varied and useful attributes. Goldenrod was used medicinally for colds, liver and kidney problems and skin aliments. The young greens were eaten and the leaves and flowers were dried to make herbal infusions and teas. Children played with the flower heads, pretending they were play whips.

The fragrance, color and form of the whole spiritual expression of Goldenrod are hopeful and strength giving beyond any I know. A single spike is sufficient to heal unbelief and melancholy.
– John Muir

Goldenrod Gall

Moth on Goldenrod

Goldenrod With
Mantis Egg Case

All goldenrods are late bloomers, flowering in midsummer into the fall, adding colorful swaths of yellow beauty and texture to fields in vacant lots and roadsides. The American Indians relied on goldenrod's flower as an indicator in their floral calendar that corn was ripening for harvest.

The most recognized goldenrod in the Northeastern regions of the United States is Canada Goldenrod *(S. Canadensis)*. The most fragrant goldenrod is Sweet Goldenrod *(S. odora)*, one of the major herbs substituted for China tea by the colonists. When crushed, its dried leaves release a sweet anise-like fragrance. Brewed in water, they make a light, flavorful drink, historically known as "Blue Mountain Tea" or "Patriot Tea."

Goldenrod also has a long history of use by native butterflies, beetles and bees. If you stop and look closely at a goldenrod plant you will be truly amazed at the activity taking place. Be on the lookout for Ambush bugs that prey on the bees, flies, wasps and butterflies feeding on the plant's nectar. They grab and inject their unsuspecting prey with a paralyzing digestive fluid and then suck out its liquified internal tissues. Ambush bugs are most easily observed in early autumn when they are fully grown. Goldenrod plants are hosts to both flies and moths, which lay their eggs in the stems and the leaves of the plants. When the larva hatches, it stops the plant's natural growth and galls are formed on the flower heads or along the stems. One of the best places to find praying mantis cases is on the dead stalks of goldenrod. They prey on the insects foraging on the late-blooming wildflower and then lay their eggs.

Goldenrod is the symbol for treasure, and legend has it that whoever carries the plant will have good fortune.

Goldenrod is easy to start from nursery plants or from root divisions obtained in the spring. It prefers sunny locations with well-drained soil and is drought tolerant, requiring little water and fertilization.

GOOD FRIENDS — OUR NATIVE TREES

Sumac *(Rhus* sp.*)*

Sumacs are small trees or shrubs characterized by attractive red seed heads, brilliant scarlet fall leaves, yellowish-white spring flowers and the habit of spreading into thickets by their suckering root system. A variety of sumacs can be found growing in clearings, vacant lots and along road-ways throughout the United States and Canada. The most familiar type growing in the Northeastern region of the U.S. is the Staghorn or Velvet Sumac *(Rhus typhina)*. This species is well known for the soft fuzzy hairs covering its terminal branches that resemble a stag's antlers before the velvet is shed. It is also called "Vinegar Tree" because of the berries' tart flavor.

Sumac

The presence of staghorn sumac is an indication of poor soil and a sig-nal to farmers to look for a site with richer soil for planting crops. Be-cause the young trees grow rapidly and easily adjust to disturbed sites, it is an important transition plant in environmental rehabilitation. Sumacs bind the soil and encourage other sun-intolerant trees to estab-lish in their shade and then die young to allow for the release and growth of the longer lived and larger trees they have sheltered.

American Indians used the berries of the sumac tree to make a refresh-ing tart drink, reminiscent of lemonade, often called "Indian lemonade". They used the roots, bark and leaves in medicinal preparations to cure a variety of ailments. Sumac leaves were used for a brown dye, the roots produced a yellow dye and the inner bark and pith of the stem were mixed with bloodroot and used for orange color. The split bark and stems of the sumac were gathered for basket making. After punching out the soft central pith, sumac branches can be used as natural taps for collecting maple sap, or making whistles and pea shooters for children.

Staghorn sumac is an attractive choice for landscape plantings around the home where spreading from root suckers will not be a problem. Its distinctive foliage, furry twigs and bright red berries make it one of the best ornamentals for natural plantings. Birds love the berries and it provides an excellent winter emergency source of food for wildlife. Honeybees are attracted to the flowers in spring. Native carpenter bees build their nest cells in the pithy centers of the tree branches. For ornamental gardens, shiny sumac *(Rhus copallina)* is highly recommended because it is long lived, and its suckering tendencies are more restrained than other sumac varieties.

ACTIVITIES AND ADVENTURES FROM
THE BACKYARD AND BEYOND

HARVEST DOLL

Cornhusk dolls are traditionally associated with the harvest time, but many other natural materials can be used to create harvest figures. Ornamental grasses are frequently used in home and public landscapes and are readily available in urban and rural neighborhoods.

Harvest Doll

Materials needed:
- Long-stemmed bundle of grass
- Small gourd
- Wire
- Raffia
- Tacky glue
- Tulip tree leaves
- Dried corn silk
- Dried black turtle beans

Gather a good size bundle of long stemmed grasses and a small gourd with the top removed. Divide the grass into three bundles; two small bundles of grass for the arms and one a bit larger, with longer stems for the main body.

Take the larger bundle of grass for the main body of the doll and push the stems snugly into the hole of the gourd where the top has been removed. The two small bundles of grass will create the doll's arms. Use wire to attach the grass bundles around the main body just below the gourd head. Split the grass body into two halves and tie at the bottom to create the doll's legs.

Glue a tulip tree leaf to the front and back of the doll to make a shirt. Tie a grass blade into a bow tie and glue on the front of the shirt. Glue dried corn silk to the top of head for hair and make a face with colored markers and dried beans.

Allium Magic Wand

Materials needed:
- Long-stemmed dried allium flower
- Paint
- Dried flowers and leaves
- Small paper doily
- Glue
- Colored ribbon

Cut a dried allium flower from the garden leaving the long stalk attached. Spray paint the flower head whatever color you prefer. Spray paint or color by hand a paper doily and glue beneath the bottom of allium head. Gather flowers and herbs from the garden associated with magical folklore and glue around the doily to frame the allium head. Wrap the stalk with colored ribbon and make a wish.

Pumpkin or Squash Stem Whistle

Materials needed:
- Pumpkin or squash leaf with stem attached
- Small knife

Cut a sturdy six- to eight-inch pumpkin or squash leaf stalk (petiole) with leaf blade attached. The leaf petiole is solid at the point where it joins the leaf blade and the rest of the leaf petiole gets hollow as it widens toward the base. Before you start handling the leaf petiole, remove the rough little prickles by gently wiping them off with a terry washcloth.

THE LEGEND OF THE CORNHUSK DOLL

At one time the cornhusk people had very beautiful faces. The Creator placed then on the earth to be companions to the children. Their task was to play and entertain the little ones. After a while though, the lovely cornhusk people became obsessed with their own beauty. They forgot their appointed task and spent most of their time gazing at their own reflections in calm pools of water. The children began to complain that the cornhusk people would not play with them. The Creator, greatly disturbed at this, took away the faces of the cornhusk people and made them into dolls.

Stem Whistle
Photo courtesy of Coleen French

Trim the leaf blade off the leaf petiole and cut a one-inch slit on one side of the petiole about one inch from the solid end of the stem. (The petiole will be hollow at this point.) Wiggle the knife a bit to make the slit wide enough to get a small amount of air through. Then blow hard through the slit in the leaf petiole to hear your whistle.

THYMELY TIPS AND SAGE ADVICE

• Some studies on bird migrations suggest that a number of birds rely on the Harvest Moon to migrate from one area to another, while others wait for this moon to begin their migration. Watch for their silhouettes as they migrate by the light of the Harvest Moon.

• The American Botanical Council, http:/abc.herbalgram.org, is a good source for reliable herbal medicine information.

• *Allergy-Free Gardening,* by Thomas Leo Orgen, is a helpful book for those people who suffer from hay fever and asthma. The book contains useful information on selecting plants and trees with low allergen ratings. He rates plants on a scale from one to ten, with one being the most allergy-free selections. Allergies to plant pollen are often seasonal. Trees typically pollinate in May, grasses can be troublesome in May and June, and ragweed becomes a problem in August and September.

• Bring your houseplants back inside when the outside temperatures begin to drop. They should be settled back inside before the temperatures dip below 55° F. Check them for insects infesting the soil or leaves a few days before you plan to bring them inside to treat the plant if needed.

• Save some summer for winter. Dry and press summer and fall flowers to use for card making and scrapbooking projects. Flower blossoms, tiny leaves, sun-bleached grasses, and fern fronds can all be pressed for future projects.

• "GreenBridges" is a native herb conservation program developed by members of The Herb Society of America. Home gardeners, schools and community organizations interested in taking part in native herb conservation can find resources and guidance on how to use native herbs in their local landscapes. The program acknowledges participants for their individual efforts. Please visit their Web site at www.herbsociety.org/greenbridges/greenbridges-initiative.

OCTOBER

One say October; one says nature lies down to sleep;

the gardener know better, and will tell you that

October is as good a month as April.

You ought to know that October is the first spring month,

the month of underground germination and sprouting,

of hidden growth, of swelling buds.

— KAREL CAPEK

Hunter's Moon — White Frost On Grass Moon — Falling Leaves Moon

This is the month of the migration of birds, of the finished harvest, of the nut-gatherings, of cider making and toward the conclusion of the change of color in the trees...Its noblest nature is a certain festive abundance for the supply of all creation.

– Leigh Hunt

Some months in the Native American calendar have several moon names, or overlapping names depending which tribe's seasonal calendar you reference, the translations they used and the weather patterns in their regions.

The beautiful changing color of autumn leaves is the result of a chemical process which takes place in a tree as the season changes from summer to winter. All during the spring and summer the leaves have been manufacturing the food necessary for the tree's growth and throughout the growing season chlorophyll is continually being produced and broken down and the leaves appear green. With the arrival of fall, as the days shorten and the temperatures begin to cool, the leaves stop their food-making process. The chlorophyll breaks down, the green color disappears and the other colors in the leaves become visible as the tree prepares for winter.

Certain colors are characteristic of particular species of trees. Oaks turn red and brown, tulip trees turn yellow, dogwood purple-red, sugar maples turn yellow-orange and red maples turn a scarlet-red. The timing of the color change varies depending on the species. Sassafras and sumac are two of the earliest to change, sometimes beginning to show color changes by mid-September. Oaks begin their color change long after many species have shed their leaves. The Falling Leaves Moon signaled it was time to finish harvesting, drying and storing winter food supplies.

The first frost usually occurs in many parts of the Northeastern regions of the United States in October. First and last frost dates have always been of great inter-

Frost on Leaves

est to farmers and gardeners. Many old timers believed there was a connection between full moons and frosts, hence the old adage *"Clear moon, frost soon."* Old weather lore suggests that an early killing frost is a sign of a bad winter.

According to woolly bear weather lore, a harsh winter is expected if many woolly bears are seen before the first frost and their black bands are wider than the rust-orange band separating the black bands. The Banded Woolly Worm is the official name of the woolly bear, the caterpillar stage of an Isabella Tiger Moth.

October was called The Hunter's Moon because it was considered the best time of year to hunt. After the leaves fell off the trees and the frost killed the vegetation it was easier to pursue and kill deer, moose, fox and wolf. Like the Harvest Moon, the Hunter's Moon was also especially bright and long in the sky, providing hunters the opportunity to stalk prey at night.

Bird and Berry Walk

The variety and colors of fall berries hanging from the trees and shrubs rival the flowers during this month of the year. The end of summer and the arrival of autumn can be tracked by simply observing the colorful changing leaves, the bright berries, and fruit of trees. Plan a visit to your community nature center or metro park for a fall bird and berry walk. Be sure to bring along a hand lens, pencils, clipboard, paper and wildflower and tree guides. Many public gardens and nature centers have areas planted with wild trees, shrubs and flowers. Find this area and make a list of the berries and their colors. Some of the berries to look for:

- Rose hips *(Rosa spp.)* – Orange-red berries
- Indian cucumber *(Medeola virginiana)* – Dark-red berries

Lets just wander
here and there
Like leaves floating in
the autumn air
And look at the
common things –
Stones on the beach –
Flowers turning
into berries
from the winds
we'll catch a bit
of that wondrous
feeling that comes
not from seeing
but from being part
of nature.
– Gwen Frostic

- Pokeweed *(Phytolacca americana)* – Dark purple berries
- Baneberry *(Actaea pachypoda)* – White berries
- Jack-In-the Pulpit *(Arisaema triphyllum)* – Scarlet-red berries
- Wild Cherries *(Prunus serotina)* – Reddish-black berries
- Elderberries *(Sambucus nigra)* – Blueish-purple berries
- Dogwood fruit *(Cornus florida)* – Red berries
- Sumac *(Rhus* sp.*)* – Fuzzy red clusters

After you have made your list, find a spot to sit quietly and watch to see who stops by to eat the berries. Make a list of the wildlife you observe and what berries they are eating. Early morning and late afternoon are prime feeding times for birds. Use your hand lens to look closely at the different types of berries.

Bird Watching

The arrival and departure of birds is an indicator of the seasons. *Journey North* has a citizen scientist project to help track hummingbird sightings in local regions as they travel back and forth to their winter grounds. Visit their interactive Web site, www.learner.org/jnorth/humm, to learn more about these incredible diminutive dynamos and how you can help them along on their cross county travels.

Leaf Art

Find a location somewhere in your backyard or beyond to collect leaves. When collecting, look for unusual shapes, different sizes, textures, colors and compound and simple leaf types. Examine leaf edges or margins with a hand lens; they can provide clues to a tree's identity. While you are walking around gathering leaves, be sure to notice the various colors of the stems and bark of trees and shrubs.

Pokeweed Berries

If in the fall of the leaf, in October many leaves wither on the boughs and hang there, it betokens a frosty winter and much snow.

– TRADITIONAL LORE

Many things can be made with dried and fresh leaves collected from the trees in your backyard and beyond. The simplest way to dry autumn leaves is to lay them in an old telephone book. Take one along on your walk and press them between the pages as you walk along.

A flower or leaf press is a useful tool for people who collect and press plants throughout the year. For directions on making a leaf press, check the internet or books at your local library. Botany students use them to make the detailed herbarium collections.

Some leaves dry better than others and keep their bright color and shape, while others just shrivel and turn brown. Experiment drying a variety of leaf types to find which dry best. Store dried leaves according to their types, color or size in large paper envelopes.

Leaf Stenciling

Materials needed:
- Poster paints
- Containers (for each color of paint)
- Tree leaves
- Paper
- Moist sponges
- Straight pins
- Toothbrush
- Popsicle stick
- Paintbrush

Sponge Stenciling

Pour the poster paints into containers. Place the leaf on the paper. Dip the sponge into the paint and pat the sponge up and down to get an even coat of paint on the sponge's surface. Holding the leaf in place with your fingers, sponge paint over the leaf's edges and onto the paper. Carefully lift the leaf to reveal the stenciled designed.

SPATTER STENCILING

Pin a leaf onto a piece of paper with straight pins. Dip a toothbrush into paint and shake off excess. Holding the toothbrush over the leaf. Rub the popsicle stick across the bristles of the brush, spattering the paint over the leaf and paper, covering the area outside the leaf edges with paint. When the paint is dry, remove pins and lift the leaf. Try spattering different colors of paint over one another to mimic Mother Nature. *Look What I Did With a Leaf,* by Mortetza E.Sohi, is a great book for ideas and tips for making collages and pressed leaf art. Check out your local library or the Internet for more craft ideas with leaves. The possibilities are endless.

Asters

Asters are the most abundant perennial fall wildflower in North America. There are over 120 types of native asters found in the United States and all have flowers arranged similar to the sunflower. Asters are also called starworts, Michaelmas daisies, or frost flowers. The word aster is Latin for star, descriptive of the star-like form of the aster's flower. Their showy flowers come in a variety of colors: white, violet, purple and lavender blue. The genus *Aster* is generally restricted to the Old World species, most of the New World Aster species have been reclassified into other genera. North American native asters are still widely referred to as asters in the horticultural circles.

Frost Aster

Wild asters can be found growing in swamps, bogs, woods and fields. Interesting comparisons can be made between their leaf size, shape and habitat. Asters growing in shady environments have large leaves to help them absorb the filtered sunlight. Asters growing in dry, sunny locations have small narrow leaves which help the plant conserve water and prevent drying in the hot sun. Asters growing in fields and meadows have medium-size leaves because they have access to both sun and moisture.

After most plants have died back in late fall, asters are the last plants left blooming before the first hard frost, and provide an important pre-winter food and nectar source for late foraging insects. Bumble bees, wasps, caterpillars, Sulfur and White butterflies are known to feed on the various parts of fall asters. The dried seed heads serve as a winter food for tree sparrows, goldfinches, wild turkeys and chipmunks.

New England Aster

The Chippewa Indians used two native asters as "hunting charms," the New England Aster *(A. novae-angliae)* and Purple-Stemmed *(A. puniceus).* The roots of these two asters were mixed with bearberry and other herbs and smoked in a pipe. The fragrant mixture was said to smell like the hooves of a deer and they were attracted to the odor. Today the Purple-Stemmed Aster is often planted in wildlife food plots to attract deer.

The aster and goldenrod have long been intimately associated as fall flowers, not only in fact but also in legend and tradition.

The Legend of the Aster and Goldenrod

Many moons ago, in a queer little hut near the edge of a pine forest beside a clear lake lived an old medicine woman. She had lived there so long no one remembered where she had come from. She was so old she was bent over, almost double. Her face was wrinkled, but her eyes were bright and noticed everything. It was said she had the power to change human beings into animals, birds or plants and that she could talk to the things that lived in the forest in their own language.

She sat by the door of her hut every day weaving mats and baskets. One day in late summer two children were seen wandering along the shore of the lake gathering flowers and tossing stones in the water. One of them

But on the hill the goldenrod, and the aster in the wood, And the yellow sunflower by the brook, in autumn beauty stood.

– Bryant

had beautiful golden hair and the other had soft, deep blue eyes that looked like stars.

They had heard the stories about the old woman and as they sat by the lake they talked of what they would choose to be, if she should try her spells upon them. Golden-hair wished to be something that would make every one who saw her happy and cheerful, while timid little Star-eye wished that she might be near her friend.

At last the sun began to sink in the west. The wind stirred among the tree tops and every now and then the acorns fell with a noise like rain-drops and the little girls became frightened. They saw the hut in the dis-tance and holding each other's hands ran toward it. As they drew near the old woman worked faster and pretended she did not see them. The children asked her, "Please can you tell us where the old woman lives who can make us whatever we wish to be?" The woman looked up and said: "Perhaps I can, what do you want of her?" "I want" said Golden Hair, "to ask her to make me something that will please everybody, and Star-eye wants to be near me." "Come in," said the woman, "and sit down. I will give you each a corn cake, and when you have eaten it we will talk about your wishes." The two little girls went into the hut and sat down to eat their cake. That was a long time ago and no has seen those children since, but the next morning there were two new wildflowers blossoming in the fields, on the prairies and on the mountain sides. One was a bright yellow plume that waved in the wind and glowed like gold in the sun-shine, and the other was a little starry purple flower. The two are never far apart and they are called the goldenrod and aster.

Adapted from Flower Lore and Legend, Katherine M. Beals, Henry Holt and Co., 1917.

Asters have been hybridized for several centuries and are valued by home gardeners for their bright fall colors and ease of cultivation. Pinching back garden variety asters in late June in northeastern regions of the U.S. extends flowering longer into fall.

GOOD FRIENDS — OUR NATIVE TREES

White Oak *(Quercus alba)*

The white oak tree belongs to the genus *Quercus*, believed to be derived from two Celtic words *quer*, fine and *cuex*, a tree. Oaks are trees of commanding strength, endurance and kingship. An oak may reach heights of more than one hundred feet and can live for centuries. Oaks were the great trees of the Druids, and have an ancient history as sacred trees.

There are about 80 oaks native to North America, with a few native to very small ranges across the United States. Oak species are commonly divided into one of two groups, the white oaks and the red oaks. They are characterized by the shape of their leaves and the time required for their fruits to mature.

The white oaks have leaves with rounded lobes, bloom in the spring and their acorns mature in the same season. The red oaks usually have sharped tipped lobes on their leaves, bloom in the spring, but their acorns do not mature until the fall of the following year.

The white oak is considered by many to be the most valuable and beautiful of our stately oak trees. It has a long history of use by the American Indians. The bark was added to their medicinal compounds to heal an amazing range of physical as well as psychological ailments. A decoction was taken to counteract the loneliness *"when your woman goes off and won't come back."* The acorns were boiled and used for soup and dried for flour. The high-grade wood was used to make tools, baskets and poles for wigwams. The colonists harvested vast tracks of white oaks for lumber to build ships and make furniture, barrels and wagon wheels.

Oak Leaves the Size
of a Mouse's Ear

The oak is the most majestic of trees. It has been represented as holding the same rank among plants of the temperate hemispheres that the lion does among the quadrupeds, and the eagle among the birds, that is to say it is the emblem of grandeur, strength, and duration; of force that resists as a lion is a force that acts.

– LOUDEN

Native Americans depended on white oak trees as important landmarks and seasonal indicators within their local regions. For planting corn, they would say *"as to the time of planting our corn; when the leaves are the size of a mouse's ear, then it is time to put the seed in the ground."*

The white oak is an impressive long-lived deciduous tree that will thrive in many locations. It is probably the finest of all oaks for landscape planting, provided the surroundings are spacious enough for it to mature to its full potential. In the early spring the tree's newly emerging leaves are tinted pale rose and covered with a soft, beautiful, silvery-white fuzz. Then by early summer its lustrous large leaves are a deep-olive green which turn a wine red color in autumn. The acorns mature in one year and will appear on the current year's growth and drop in late September.

A white oak tree is a good place to observe the interdependence of plant and animal life. Its leaves, bark, acorns and roots are eaten or utilized by a variety of birds, insects and animals. Birds that nest in oak tree cavities include chickadees, woodpeckers, owls, and bluebirds. Tree crickets, leafhoppers, oak-bark beetles and leaf miners eat the leaves and provide food for the birds living in the tree. Oaks support more species of butterflies and moths than any other plant genus. Squirrels, blue jays and wood mice eat the acorns. Red-tailed hawks perch in oak trees, hidden by the leaves waiting, for mice to come and eat the acorns.

Growing oaks from seed can be an interesting, informative and inexpensive gardening activity for homeowners and community organizations.

White oak acorns fall before red acorns. The bad seeds usually drop from the tree before the good seed, so wait until most of the acorns have fallen. They need to be collected promptly because they start to germinate days after they fall from the tree.

White oak acorns do best when planted where you want them to permanently grow. They start to root right away, but the top will not emerge until spring. They develop a long tap root, so if you decide to transplant them, make sure to do it before the tree is over a foot or two tall.

Acorns can be started indoors. A pot at least eight-inches deep is recommended to accommodate the long tap root. Fill with a medium textured, moderately absorbent soil mixture and plant the acorn an inch deep. Place the pots in a south sunny facing window and keep evenly watered. Move the plant outside as soon as the weather permits.

ACTIVITIES AND ADVENTURES FROM
THE BACKYARD AND BEYOND

MINIATURE WILDLIFE GARDEN

Lilliputian landscape arrangements are fun to create using dried flowers, herbs, acorns and seed pods. Most plant materials need time to dry or be preserved properly. Plan ahead and gather materials throughout the cycle of the seasons.

Materials needed:

- Green spray paint
- Goldenrod flower
- One cattail with male blossoms ($3^1/_2$ " long and $1/_2$ " wide)
- Extra thick tacky glue
- Leaves
- Tiny twigs
- Small seeds
- A tiny tuff of milkweed fluff
- $3^1/_2$ " round tree cookie
- Dried moss, herbs, acorns

Wildlife Garden

Spray the goldenrod flower head green and lay aside to dry. Carefully remove the male blossom from the top of the cattail body. Then cut the male blossom in four even sections to make legs for the deer. Bend the thicker end of the cattail up at a slight angle to form the head. Turn the cattail over and carefully glue the four legs to the bottom of the deer's body. Let set until dry.

Glue two tiny brown leaves and a wee-branched twig to the top of the deer head to make ears and antlers. Glue a small, round brown seed to the end of its face to make a nose. Then glue the bit of milkweed fluff to backend of the deer to make its white tail. Cover the flat side of the tree cookie with dried moss and glue the goldenrod stem down through the moss for a pine tree. Arrange the rest of the dried flowers, herbs and acorns to create pleasing habitat for the deer. Glue the deer down in front of the pine tree.

NATURE MOBILE

This is a good project to do with a group. Working together and helping one another build and balance a mobile is a simple way to demonstrate the fragile balance of nature.

Start by assembling an assortment of natural objects of varying weights and sizes to use on your mobile. Not all of them will be used on one mobile, but you will need a large assortment of objects so you can make substitutions as you work on balance.

Nature Mobile

Materials needed:
- Wire cutter
- 18 gauge wire
- One small rock
- Two small tree branches, about 12" long and $1/2$ " to 1" thick.
- Pine cones, seed pods, bark, dried flower heads, leaves, small rocks and nuts can all be used.
- Heavy weight fish line
- Scissors

Cut an 18-inch piece of wire and wrap one end of the wire around the small rock several times, leaving a long piece of the wire. Tie the other end of the wire with the rock attached to the end of one of the twigs.

Pick another object about the same weight as the rock, wrap and tie it with the heavy fish line and attach it to the other end of the twig. You will now have a twig with one object attached by fish line and a rock attached with 18-gauge wire.

Tie a hanger to the middle of this branch using the wire. Tie the fish line to the wire hanger and attach it to the top branch of the mobile. Attach a wire hook on the center of this branch to hang the mobile. Then tie a piece of fish line to the middle of the wire hanger to hang the finished mobile.

Tie a piece of fish line to the other nature objects. Then tie the other end of each string onto branches. Adjust the length and placement of each string until all the objects are evenly balanced.

ACORNICLE

Children have played with parts of plants since the beginning of time. This simple game is a fun way to discover the differences between oak acorns. The object of the game is to get five acorns of one kind in a row, in five successive squares. This game may also be played with leaves or twigs.

Each player names his kind of acorn and collects fifty acorns of that type of acorn. On the ground or on a large sheet of paper mark off 200 squares. A player is chosen to start the game off and he puts the first acorn in any square. Each player after him takes a turn placing one at a time in an empty square. Opponents may be blocked by placing an acorn at the end of his row. The first one to get five in a row wins.

Adapted from *Nature Recreation*, by William Gould Vinal.

THYMELY TIPS AND SAGE ADVICE

• Now is the time of year to plant tulips, Alliums, daffodils and other spring-flowering bulbs. Bulbs can be planted any place in your yard and garden with healthy soil and good drainage. Prepare the beds by loosening the soil to a depth of six to eight inches. Planting depths depend on the size of the bulbs and can be found on the packages. Mix a little fertilizer into the soil in the bulb hole and plant the bulbs with their pointed side up.

Lasagna Gardening

Lasagna gardening or sheet composting is an easy, no-till organic method of converting areas of grass or dense weeds into new planting beds with very little labor. Fall is an excellent time to sheet compost because the materials have time to break down slowly over winter and the site is ready for planting in the spring.

Follow these few simple steps:

Preparing the site — don't worry about removing existing lawn or weeds. Mow or cut down the grass or weeds in the location you would like to establish a new bed or expand a garden. There is no one way to sheet mulch, so you can use whatever organic materials that are available to you. The basic technique involves placing alternate layers of carbon or brown materials that are dead and dry, and nitrogen materials which are green, wet or fresh directly onto the ground. Dry leaves, straw, sawdust and newspapers are good brown sources, non-animal food scraps, composted animal manure, soybean meal and green grass clippings are good green sources. The carbon or brown layers should be about twice as deep as your green or nitrogen layers.

Layering, cover the ground with six to eight overlapping layers of wet newspaper or cardboard. Then add a thick layer of grass clippings, vegetable scraps or green weeds, cover that layer with another layer of brown material — straw, shredded paper or dry leaves. Keep alternating layers until you have a two-foot tall, layered bed.

• Check out the November 2010 issue of *National Geographic*. It contains a great article entitled *Mysteries of Great Migrations; What Guides Them Into the Unknown?* A free map is included with the copy of the magazine and shows the patterns, habits and routes of migrating insects, birds, reptiles and amphibians across the world. It is an excellent and inspiring resource revealing the magic and mysteries of Mother Nature's natural laws and a great teaching tool for children.

NOVEMBER

November comes, And November goes,

With the last red berries, And the first white snows.

With night coming early, And dawn coming late,

And ice in the bucket, And frost by the gate.

The fires burn, And the kettles sing,

And earth sinks to rest until next spring.

– Elizabeth Coatsworth

Beaver Moon — Falling Snow Moon

Beavers have been an integral part of the European and North American ecosystems for thousands of years. The Canadian beaver is North America's largest rodent and spends most of its life in the water. There are more beavers in the United States and Canada than anywhere else in the world.

Beavers live in streams, rivers and freshwater lakes near woodlands. Largely instinct driven, beavers are known for their industrious, efficient behavior, skill at cutting down trees and building dams. For this reason, we often call a hard working person an "eager beaver" or say he is "busy as a beaver."

Beaver photo courtesy of Demian J. Betz

Canadian beavers were one of the most aggressively hunted animals in North America from the 1600s through the 1800s. Historically it was prized for its meat, its thick coat and for the oil known as castoreum that is secreted by glands beneath its tail. Many medicinal properties and special powers were attributed to castoreum. The name comes from the Greek *Kastor*, meaning beaver, as does the animal's scientific name, *Castor canadensis.* The oil was used as an aphrodisiac, to treat epilepsy and a base for perfumes. Castoreum oil contains high concentrations of salicylic acid, from which aspirin was developed. Found in the inner bark of willow trees, it is one of the beaver's favorite foods. Trading companies shipped beaver furs throughout the world to be made into hats and coats. Hunters killed so many beavers that by the late 1800s beavers were almost extinct. The Canadian and U.S. governments passed laws to protect the animal, and today, like other animals, beavers can only be hunted at certain times of the year.

American Indians recognized the beaver's beneficial influence on the health of ecosystems along the banks of North American waterways, wetlands and lakes and sometimes referred to them as the *"earth's kidneys."* The animal was admired and respected for its strong will, creativity, wisdom and cooperative spirit.

Beavers selectively harvest the trees they prefer as foods, primarily alders, aspens, cottonwoods and willows. They also like to eat serviceberry, ferns and skunk cabbage.

The European Beaver, *Castor fiber,* seldom builds dams and the North American Beaver usually does. Beaver dams are made of logs, branches and rocks plastered together with mud. Mud and stones are used for the base of the dam. Beavers have been known to keep their dams in good condition for many years.

Beaver dams, like sugar camps, were tended to by the American Indian family living near the beaver colony and were passed down from father to son. Hunting parties would never consider hunting beaver at a dam tended to by another family. Beaver traps were set during the month of November before the waters froze, so furs and meat would be available for the cold months ahead; thus the name, Full Beaver Moon. Another interpretation suggests that the name comes from the fact that beavers are actively preparing for winter at this time. Other tribes called this The Falling Snow Moon because this was the time of the first measurable snowfall in their regions.

Indian Summer

The term "Indian Summer" dates back to the eighteenth century, originating in the Northeastern and Great Plains regions of the United States. It can be defined as "any spell" of warm, quiet, hazy weather that often occurs in late October or early November. The following explanation about the possible origins of "Indian Summer" was taken from *The Century Book of Facts* published in 1902.

No warmth,
no cheerfulness,
no healthful ease,
No comfortable feel
in any member –
No shade, no shine,
no butterflies, no bees,
No fruits, no flowers,
no leaves, no birds –
November!"
– THOMAS HOOD

Scientists differ regarding the cause of this phenomenon which is particular to North America and certain parts of Central Europe. A change in the condition of the upper strata of the atmosphere, confining the radiating heat-rays in the lower strata, is generally held to be the true explanation. A theory to account for the smokey appearance, which appears to be plausible, is that due to the decay or slow chemical combustion of the leaves, grass, and other vegetable matter under the action of the frost and sun. It was to forest and prairie fires kindled by the Indians that the early settlers attributed the smokey appearance of the season. Hence the name "Indian Summer."

– *The Century Book of Facts*, p.331. Edited by Henry W. Ruoff,1902.

Any place that has a clearly defined winter season can experience an "Indian Summer." It is the final spell of warm weather in the yearly cycle of the seasons before the ground freezes solid and heavy snow starts to fall, ushering in the time of Long Nights Moon.

Rock Hunting and Collecting

November is a great month to make a rock collection. The frost has killed all the leafy green plant material in the backyard and beyond. It is a perfect time of the year to get out and hunt for rocks and see rock formations before the snow arrives to cover the ground.

Mortar and Pestle

Native Americans used rocks for a variety of purposes. They fashioned heads from stone to shaft their arrows for hunting as well as hatchets for cutting down trees. Rocks were used as mortars and pestles for grinding maize, herbs, seeds, berries and nuts. American Indian rock art marks the walls of canyons, rock outcroppings and caves across North America. Petroglyphs are images carved or pecked on stone surfaces and pictographs are paintings on rocks. Pictures of symbols that serve as important records and provide valuable clues of their visions, legends and regional histories.

Since humans first roamed the earth, cairns, shaped rock piles, have been used as markers on a mountain summit, wilderness path or commemorative site. Some were used as navigational aids and others to mark a shallow body of water, such as a river or stream where a crossing could safely be made on foot or horseback. Long ago the "language" of cairns was well known to backwoods travelers, evoking a sense of safety, hope and friendship, imbued with well wishes to travelers.

Petroforms are boulder or rock mosaics that were created by lining up large rocks on open ground in patterns. They were built and used by Native Americans for astronomical, religious, sacred, healing and teaching purposes. Some of North American petroform shapes are over 2,500 years old. Many mirror the night sky and the patterns of the stars. The Ojibway used stone petroforms to make medicine wheels representing the Seven Sacred Directions, which included the Four Sacred Directions — the East, West, North and South — Mother Earth below, Father Sky in the upper realm, and self. To learn more about this Indigenous knowledge and philosophy, please visit the online Web site to view the program produced by the Department of Canadian Heritage, http://www.fourdirectionsteachings.com.

Rocks fall into three categories according to the way they were formed.

Igneous rocks

At one time these were molten liquids that became solid rock as the earth cooled years ago. Examples are granite, pumice, basalt and obsidian.

Sedimentary rocks

These rocks were formed by great pressure on top of silt, sand and other small particles squashed down together to form solid rock. Shale, sandstone and limestone are a few examples of sedimentary rock. Some-

AUTUMN
CONSTELLATIONS
EVERY CHILD
SHOULD KNOW

Taurus – The Bull

Ursa Minor –
The Little Bear

Ursa Major – The Big Bear

Aquarius – The Beaver

Pisces – The Fish

times you can see layers of soil particles in this rock.

Metamorphic rocks

Metamorphic rocks are sedimentary rocks formed by being subjected to heat and pressure. Slate and marble are types of metamorphic rock.

Rocks are easy to find; begin looking in your backyard and expand into your neighborhood. If you collect on another person's property, obtain permission before you start. Never collect in a state or national park.

Look for rocks with fossils, interesting patterns and textures. Rocks that have a white stripe around them are believed to bring good luck to the person who finds one. Rocks can be sorted according to color, weight, shape, degree of hardness and type of rock. Look for rocks shaped like animals and try cracking an uninteresting rock open to see what it looks like inside.

Turtle Rock
Photo courtesy of
Susan Belsinger

Rocks are very useful for craft projects. Paint a design or interesting texture on the surface of a nicely shaped rock to use as a garden decoration. Make an arrangement of small rocks to represent a bird, snake or turtle some place in your yard.

Everybody Needs a Rock, New York: Antheneum Books, 1976 by Byrd Taylor, is a lovely book that describes exactly how to go about finding your very own personal rock. It is a great resource for teaching observational skills on nature walks or backyard adventures.

Field Horsetail *(Equisetum arvense)*

Field horsetail is a primitive plant whose ancestry dates back to the days of the dinosaurs. *Equisetum*, the only genus in the Equisetaceae, or horsetail family, has 25 or so remaining species left in existence. The branching stems of many horsetail species supposedly resemble a horse's tail, inspired the genus name of *Equisetum*. In Latin *equus* means horse and *seta* means bristle. *Arvense* means field. Native to the northern regions of the United States and Canada, it can be found growing in dry sandy, poorly-drained soil along roadsides, railroads, embankments and waste places.

Horsetail is a non-flowering herbaceous perennial. Its structure is somewhat reminiscent of bamboo, with hollow, jointed stems encircled at each node with inconspicuous whorls of leaves. They are similar to ferns in their reproductive habits. Early in the spring, before most other plants emerge, single fertile shoots appear and release spores that are spread by the wind. After the spores have been released, the shoots wither and multi-branched fronds appear from an underground creeping rhizome, producing new stems well into late autumn. It is a very interesting plant to closely observe with a hand lens.

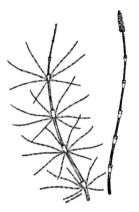

Horsetail

Field horsetail's common names provide clues to the plant's many herbal virtues. Pewter plant, shave grass, bottle-brush, and scouring rush describe its historical use by the colonists. The plant contains a large amount of abrasive silica, the stiff stalks were bundled together and used to scour pots, shine various metals such as pewter and smooth wooden surfaces. American Indians used the plant to polish arrowheads.

Horsetail was used externally to heal wounds and stop bleeding. The Romans called it "the hair of the earth" and created a healthful restorative tonic from the stalks. Old timers simmered horsetail in water to make an organic spray to treat fungal disease on plants. Horsetail extracts are used in a variety of consumer products such as shampoos, hair conditioners, skin cleansers and nail strengtheners.

Cranberry and Horsetail Necklace

The Indians used a variety of plant parts to make necklaces and personal adornments. Cranberries and hollow horsetail stems are great materials for making a nature necklace because they are both easily

Cranberry and
Horsetail Necklace

Dull November
brings the blast,
Then the leaves are
whirling fast.
– SARA COLERIDGE

obtained and dry well. In early fall harvest some long stalks of horsetail, cut it into varying lengths and dry on a screen. Fresh cranberries can be purchased at your local grocery store this time of year and are easy to string because they have no pits.

Materials needed:
- Needle and dental floss
- Scissors
- Package fresh cranberries and horsetail sections

Thread needle with a long length of floss. String cranberries onto dental floss using the needle. Alternate the berries with the hollow pieces of horsetail. Knot both ends of the string of horsetail and cranberries to form a necklace.

GOOD FRIENDS — OUR NATIVE TREES

Witch Hazel *(Hamamelis virginiana)*

The common witch hazel is native to eastern and central North America. It can be found growing in woods, along streams and north facing slopes in nearly every state east of the Mississippi River. It is also found in Canada along the Great Lakes and the St. Lawrence River. Spotted alder, snapping hazelnut and winter bloom are a few of the tree's common names.

Witch hazel is a deciduous multiple-stemmed tree or shrub that may grow 15 to 20 feet high and 25 feet across. Its smooth bark is grayish-brown and the tree's long branches are usually forked and twisted. The oval, dark green leaves are slightly fragrant and turn yellow in autumn.

Witch hazel has some interesting and curious habits that set it apart from other trees and shrubs. It is the most inconspicuous tree in the woods during the spring season, then by late fall when all the other trees are preparing for winter, its thin twisted branches burst forth blooming in shocks of spidery yellow flowers. The fruit, which takes a year to mature, is a woody capsule containing two shiny, hard black seeds. When ripe, the seed capsules explode, shooting the seeds up to ten feet away. This is Mother Nature's way of helping the heavy seed find a place to germinate at a distance from the mother plant, thus eliminating competition for its moisture and soil nutrients. If you bring a few fruiting branches of witch hazel indoors in late November, the warmth of the house will set off last year's capsules, catapulting seeds all over the room.

American Indians used witch hazel for a multitude of medicinal treatments and taught the European settlers about the tree's many practical attributes. The leaves, twigs and bark were used as an astringent, tonic and sedative. They made a poultice out of the leaves and small twigs to soothe skin inflammations and insect bites.

Witch hazel extracts have been used in skin creams, lotions and ointments since the 1840s. Over one-million gallons of witch hazel are purchased each year in the United States and Canada.

The tree's crotched sticks were used as divining rods to reveal the location of hidden springs of water.

Witch hazel prefers well-drained loamy soils. It can tolerate full sun to shade which makes it a good choice for a shady backyard. It is an easily grown tree, with over 100 cultivars of both native and non-native types available as landscape plants in nurseries across the county. Wild turkey and ruffed grouse consume the seeds, rabbits and squirrels eat the bark, leaves and seeds, and white-tailed deer eat the twigs, buds and leaves.

Witch Hazel

Through the gray and
somber wood
Against the dusk
of fir and pine
Last of their
floral sisterhood
The hazel's yellow
blossoms shine.
– JOHN G. WHITTIER

ACTIVITIES AND ADVENTURES FROM
THE BACKYARD AND BEYOND

SAND PAINTING

Sand paintings are pictures made of bits of crushed rocks and sand.

Materials needed:
- Dry sand
- Dry powdered tempera paint
- Zippered plastic bags
- Pans or trays (for each color of sand)
- Cardboard or heavy tagboard
- Pencils
- White glue

1. To prepare colored sand, mix dry tempera paint with sand in zippered plastic bags and blend well.

2. Place the colored sand in different pans.

3. Work out a design and draw it on cardboard.

4. Outline a small area of the drawing with white glue. Drop the appropriate color of sand over the glued area and tap to let the unglued grains fall back into the container.

5. Follow step 4 until the painting is completed.

OBSERVATION WINDOW

With temperatures dropping below freezing and shortened daylight, it may be hard to get outside and enjoy nature during the cold winter months. While venturing outdoors cannot be replaced, it can be supplemented in fun and simple ways for all ages. Set up an indoor bird-watching station in your home. Place a chair and basket containing a clipboard, pencils, art materials, binoculars and a simple bird book in the best location in the house for viewing wildlife at your backyard feeding station.

You can provide special feeders and food to attract specific types of birds and animals to your yard. Ground feeders tend to be used by squirrels, chipmunks, mourn-

The Sky
The Sun,
The Stars,
The Moon,
The Rocks,
The Earth,
The Water,
The Plants ,
The Trees,
Have always been
the teachers
of the Anishinabek.

ing doves, sparrows, cardinals and juncos. Raised feeders attract blue jays, purple finches and grosbeaks. Tube feeders are frequented by chickadees, finches and an occasional blue jay. Suet feeders will bring in woodpeckers, grackles and starlings looking for a high-energy snack. The size of the opening, the spacing of the perches and the type of seed you supply will also dictate which birds use your feeders.

Project Feeder Watch is a citizen science project that focuses on winter season bird identification. Visit their Web site, www.birdsleuth.net, and discover a wealth of resources and tips to enhance your winter bird-watching activities.

THYMELY TIPS AND SAGE ADVICE

"At this season, every sense is keened and quickened the sultry sameness of dog days of August is over and done. November is all contrasts, the sun and cloud shadow, the crackle of leaves in the forest, the sharp smell of smoke in the evening damp, black limbs on white moons, the first icy touch of snow on un-mittened hands. Certainly if we do nothing else we should see now that all is snug and ready for winter. If winter is just around the corner, let it come."

– The Old Farmer's Almanac, 1948

AUTUMN SOUNDS

Crunchy fruit

Farewell bird songs

Dropping nuts

Crispy leaves

PATTERNS

Rock formations

Bright berries

Leaf veins

Frost

Flames

Bulky clouds

Winter

Each season has
its own wonder,
its own special place
and a purpose
in the pattern
of creation.

– Cicero

Allegory of Winter and Summer

A man from the North, gray-haired, leaning on his staff, went roving over all countries. Looking around him one day, after having travelled without any intermission for four moons, he sought out a spot on which to recline and rest himself. He had not been long seated, before he saw before him a young man, very beautiful in his appearance, with red cheeks, sparkling eyes, and his hair covered with flowers; and from between his lips he blew a breath that was as sweet as the wild rose.

Said the old man to him, as he leaned upon his staff, his white beard reaching down upon his breast, "Let us repose here awhile, and converse a little. But, first we will build up a fire, and we will bring together much wood, for it will be needed to keep us warm."

The fire was made, and they took their seats by it, and began to converse, each telling the other where he came from, and what had befallen him by the way. Presently the young man felt cold. He looked round him to see what had produced this change, and pressed his hands against his cheeks to keep them warm.

The old man spoke and said, "When I wish to cross a river, I breathe upon it and make it hard, and walk over upon its surface. I have only to speak, and bid the waters be still, and touch them with my finger, and they become hard as stone. The tread of my foot makes soft things hard — and my power is boundless."

The young man, feeling every moment still colder, and growing tired of the old man's boasting, and the morning being nigh, as he perceived by the reddening east, thus began —"Now, my father, I wish to speak."

"Speak," said the old man, "my ear, though it be old, is open — it can hear."

"Then," said the young man, "I also go over all the earth. I have seen it covered with snow, and the waters I have seen hard as stone; but I have only passed over them, and the snow had melted; the mountain streams have began to flow, the rivers to move, the ice to melt: the earth has become green under my tread, the flowers blossomed, the birds were joyful, and all the power of which you boast vanished away!"

The old man drew a deep sigh, and shaking his head, he said, "I know thee, thou art Summer!"

"True," said the young man, "and here behold my head — see it crowned with flowers! And, my cheeks how they bloom — come near and touch me. Thou art Winter! I know thy power is great; but, father, thou darest not come to my country — thy beard would fall off, and all thy strength would fail, and thou wouldst die!"

The old man felt this truth; for before the morning was come, he was seen vanishing away: but each, before they parted, expressed a hope that they might meet again before many moons.

"The Allegory of Winter and Summer," *Winter Studies and Summer Rambles in Canada*, Anna Jameson, p. 366-368. This tale was told by O-sha-gush-ko-na-qua, daughter of Chippewa Chief Waub-Ojeeg to Anna Jameson while visiting during 1837.

DECEMBER

A trip to the woods in winter

is a revelation to some.

We come upon groups of evergreen ferns

and never appreciate them more.

Here, the voice of a winter bird

sounds an innocent note.

If it is night there is the clear moon

which transforms each bush

into a priceless silhouette.

— ALFRED CARL HOTTES, 1949

Cold Moon — Long-Nights Moon — Falling-Snow Moon

Full Cold Moon announced the arrival of winter, the coldest season of the year between autumn and spring, marked by the shortest days and the longest nights. Even though the daylight hours are less in the winter months, the earth is actually closer to the sun. The earth's revolution around the sun and the tilt of its axis are the reasons for the changing seasons. The earth's position is constantly shifting; thus, month by month, the constellations are also changing overhead. The stars we see in the winter sky are not necessarily the ones we will see during other seasons of the year.

Full Long-Nights Moon describes the sky in December, when the moon appears earlier and rides lower across the evening horizon and the stars put on the brightest show of the year.

The name Full Falling-Snow Moon signaled that it was time to begin observing the ways of snow and ice.

During the winter months, American Indian tribes would split up into extended family groups, dispersing to campsites in more sheltered locations. Each season had its own particular tasks, activities and ceremonies. After the women and children were settled in their winter lodgings, the men and boys would head out to their winter hunting grounds. Wintertime was spent mending tools and preserving furs.

For evening fun and entertainment, young women seated around the fire would make birch bark pictures. Using their eye teeth they would bite figures on the soft inner layers of the inner bark of the *Betula papyrifera* or paper birch. The designs were patterned after local flowers, leaves, animals or landmarks. Some pictures were kept as mementos, but most were used as beadwork patterns for clothing and headbands. Snow Snakes was a favorite winter game played by the children, who would take turns shoving a large stick along a path of packed snow to see whose "snake" would go the fastest. Making snowshoes and tobogganing were some of their other winter pastimes.

The Indians watched "nature's timekeepers" the sun, moon and stars to track the passage of time on a daily, monthly and yearly basis. Winter solstice occurs on the shortest day of the year and marks the beginning of winter; the silent hermetic period in the year's cycle as the earth rests before awakening and growing anew. On the calendar it falls between December 20 and 22 in the Northern Hemisphere. The word solstice means standing-still-sun, a time of stillness before the sun's strength builds and the days grow longer.

Cultures the world over have long recognized and celebrated winter solstice as a turning point in the cycle of the seasons. It is a time to appreciate the return of light and the time to begin looking forward to the coming of spring. For American Indians, time was circular rather than linear, a continual cycle of life, birth, death and rebirth. The circle has neither beginning nor end and it symbolizes the sun, the moon, and the calendar year. As part of the universal circle, the tribe encircles the family and at the circle's center stands the individual who also passes through the circle of life, changing with each season.

The life of a man is a circle from childhood to childhood, and so it is in everything where power moves. Our teepees were round like the nests of birds, and these were always set in a circle, the nations hoop, a nest of many nests, where the Great spirit meant for us to hatch our children.

— BLACK ELK
LAKOTA SIOUX

Moonlight Adventures

The winter months are an ideal time of the year to introduce children to sky watching. Because of the shorter daylight hours, it is dark long before young people have to be in bed. Before or after dinner, spending a bit of time together as a family viewing the moon or learning a new constellation is a pleasant way to close the day. The constantly changing light and shape of the moon plays a significant role in the appearance of the night sky and the stars rise four minutes earlier every night, so no two

Twinkle, twinkle,
little star
How I wonder
what you are!
Up above the
world so high,
Like a diamond
in the sky.
When the glorious
sun is set
When the grass
with dew is wet
Then you show
your little light,
Twinkle, twinkle,
all the night.
In the dark-blue sky
you keep,
And often through my
curtains you peep:
For you never
shut your eye
Till the sun is in the sky.
As your bright
and tiny spark
lights the traveler
in the dark ,
Though I know not
what you are,
Twinkle, twinkle,
little star!

nights are ever the same. It's easiest to begin observing the stars when the moon is in its last quarter, so the stars are not diminished by the light of the moon.

Ancient civilizations found that by connecting the stars as if they were dots, patterns appeared resembling lions, bears, hunters and other fantastic creatures. It takes patience, a lot of imagination and a good sky map to locate constellations and other interesting curiosities of the ever-changing night sky.

The Abrams Planetarium produces *Sky Calendar/Evening Skies*, a detailed daily guide to the evening skies covering the entire continental United States. It gives a calendar sheet for every month showing the moon, planets and significant sky events to watch for. The reverse side consists of a simplified star map of the month's evening sky. A subscription is available for a minimal cost, and will provide hours of year-round, nocturnal family fun. For more information and ordering details visit their Web site, http://www.pa.msu.edu/abrams/SkyWatchersPage/Index.

Observe and sketch the position and size of the moon at the same time each night until the moon is fifteen days old. Label each sketch with the moon's age. Use a pair of binoculars or a telescope to get a better view of the moon's surface. Many people can see a face when they look at a full moon and cultures from all over the world have interesting legends to explain how it got there.

According to one legend, the man in the moon was a gardener when he lived on earth. Long ago, one Christmas Eve he was hungry and in a rush to get home, when he passed his neighbor's vegetable garden, he stuffed several cabbages into his basket for dinner. As he hurried up the walk to his house with the basket of cabbages, a voice called out, *"because thou has stolen on the holy night, the moon shall be thy home forever."* Supposedly, he is still up there to this day, the exiled Man in the Moon.

When the moon is full, sketch your impression of what the face in the moon looks like on a round flat piece of paper. Then write your own legend about how he or she ended up there.

Holiday Plant Traditions

The ancient custom of decorating our homes with evergreens during winter holidays dates to antiquity. When cold winter weather killed most plants, the evergreens remained, symbolic of summer and life. Long before the Christian era, evergreen boughs, wreaths, garlands and trees were used to symbolize eternal life, protection and prosperity. In the Celtic and Nordic lands they were highly esteemed and it was believed such plants as the pine, spruce, fir, mistletoe, ivy and juniper shielded them from evil spirits during the dark winter months as they awaited the return of light and warmth. The Romans lavishly decorated their homes and temples with evergreen boughs and laurels in celebration of *Saturnalia*, honoring Saturn the god of agriculture.

Rosemary is said to have been one of the plants to give shelter to the holy family when they were being pursued by Herod's soldiers. As they rested for the night, Mary threw her cloak of blue cloth over a bush of Rosemary covered with tiny white blossoms and when she removed it the next morning the white flowers had turned to blue.

Holly, according to Pliny the Elder, was a plant of many virtues, with a host of supernatural qualities. It was planted near the house for protection against lightning and witchcraft. People believed the flowers would cause water to freeze, and when a branch of holly was thrown at an animal, it was obliged to lie down beside it. The first person to bring holly into the house during the holiday season, husband or wife, would rule for the year. A British tradition was to place a piece of holly in each bee-

Winter holly and ivy
So green and gay.
We deck our houses
As fresh as the day.

With bays and rosemary
and laurel complete,
and everyone now, is a
king in conceit.
Poor Robin's
Almanac, 1695

Holly

hive at holiday time. This custom honored a legend that held when the Baby Jesus was born in Bethlehem, the bees sang a song in his honor, a song that they have been humming ever since.

The custom of using holly at Christmas was brought to the United States by the English settlers. The native American holly, *Ilex opaca,* is probably the best known and widely grown holly in the United States. Because of its similarity to English holly, *I. aquifolium,* English settlers used American holly for holiday decorating. Today, the availability of hundreds of named cultivars, including one called 'Merry Christmas', attests to the American holly's enduring popularity. Holly and ivy have always been used together in decorations. Holly is thought to be more masculine with its thorny leaves and hard edges, while ivy appears more gentle and feminine with its softly-edged leaves and clinging habit. In the language of flowers, holly represents "foresight" and ivy stands for "friendship." Because of their glossy, dark green leaves, grayish bark and clusters of bright red berries, evergreen hollies offer four seasons of landscape appeal.

Many charming legends exist about the origin of the Christmas tree. In one version, as Christianity spread across Northern Europe, Faith, Hope and Charity were sent down from heaven to locate and light a tree that was as high as hope, as great as love, as sweet as charity and with the sign of the cross on every bough. Their search ended when they found a fir, lit by the radiance of the stars, it became the first Christmas tree. The triangular shape of the fir has also been used to describe the Holy Trinity of God. The Christmas tree custom was brought to America by the Hessian mercenaries during revolutionary times and early German immigrants who settled in Pennsylvania around 1746.

Many of our well-known evergreen conifer (cone-bearing) trees, cedars, firs, hemlocks, larches, spruces and pines are classed together

A simple decorative mistletoe ball can be made by sticking short sprigs of holly, fragrant herbs and mistletoe into an apple, entirely covering the surface and then trimming the plant material so the ball is perfectly round. A greening pin attached to the top of the ball tied with a ribbon can be used to hang the ball.

Romans considered mistletoe a symbol of hope and peace; therefore, when enemies met beneath it, they would lay aside their differences and declare a truce until the following day.

Several species of mistletoe are native to America. Most of the commercial mistletoe sold in the US comes from California.

under *Pinaceae*, the pine family. These trees are covered with tough, waxy needles which persist on the tree all year long. This keeps them from losing too much water in the winter, when the ground is frozen or dry, so they do not need to drop. Most needle-leaved trees have cones to hold their seeds. Pines, spruce, hemlock and fir trees all have cones. There are approximately 33 million real Christmas trees sold in North America each year. The spruce, fir and pine are the most popular choices for Christmas trees. December is a perfect time of the year to learn how to identify narrow-leaved evergreens, because you can find a large collection of different tree types all in one location — your local Christmas tree farm or garden nursery!

Pine *(Pinus sp.)*

Pines have long, thin needles that come fastened together at the bottom in little bundles of two, three or five. Each pine has its own special number of needles in a bundle. When you see an evergreen with its needles fastened in bundles you can be sure it's a pine. The number of needles in the bundle is an important clue for telling what kind of pine the tree is. Cones are another clue, no two pines have cones exactly alike. The Scotch Pine has two needles per bundle and the Pitch Pine is the only three-needled pine in the Northeastern states. White Pine has needles three to five inches long in bundles of five.

Spruce *(Picea sp.)*

Spruce trees have stiff, pointed, four-sided needles spirally arranged and attached by tiny stems on the twigs. It can be recognized, even at night, by the feel of its twigs when the leaves have fallen, leaving rough pegs where the needles were attached. The needles fall off the tree easily making it a poor choice for an indoor Christmas tree. Spruce cones al-

Go to the winter woods: listen there, look, watch, and "the dead months" will give you a subtler secret than any you have yet found in the forest.

– Fiona Macleod
Where the Forest Murmurs

ways hang down from the branches when full grown.

Fir *(Abies* sp.*)*

Fir trees have fragrant, flat needles with blunt tips, also spirally arranged but softer than those of the spruce. Their needles sit right on the twigs. When they fall they leave the twigs clean and smooth, but tiny round scars show where the needles once grew. Fir cones stand straight up on the top sides of the branches.

GOOD FRIENDS — OUR NATIVE TREES

Eastern Red Cedar *(Juniperus virginiana)*

The red cedar is an evergreen aromatic species of juniper native to eastern North America. It's a hardy tree that can be found growing on woodland edges, old fields and disturbed lands throughout the United States. Colonists noticed the red redar growing at Roanoke Island, Virginia in 1564 and immediately began harvesting it for building furniture, fence rails and log cabins. Cedar heartwood was once the main source of wood for pencils.

American Indians also used the tree for building furniture and wigwams. Cedar boughs were put on teepee poles for protection to ward off lightning. The red aromatic heartwood was used to make "love" flutes and hunting bows. Fragrant leaves of the juniper were made into incense and insecticide. The bark made a dark mahogany-colored dye. The berries, leaves and twigs were used in medicine and for flavoring food. Because of its resistance to rot, juniper wood poles were used to stake out mutually agreed upon tribal hunting grounds.

Legends and superstitions surrounding the juniper tree date back to ancient times. It was planted beside the front door of new homes to welcome guests and provide protection from witches. Before entering a home with a juniper planted beside the door, according to their own law, witches had to count every last needle. If they made an error they had to begin counting again; in doing so they became weary and would depart in despair. juniper has long been known in plant folklore as the tree of sanctuary, for it provides safe shelter for small animals and birds seeking a safe haven from hunters and cold weather. Juniper berries have been used for centuries to flavor wild game the world over; thus it seasons the very animals that seek shelter within its fragrant boughs. It was the favorite of Saint Francis, the patron saint of animals and ecology. The fragrance of burning juniper was thought to promote a restful sleep and burning the branches was a popular practice in Britain in times past.

The Eastern Red Cedar is tolerant of many soils and difficult locations. It was planted during the dust bowl drought in the 1930s to help control soil erosion in the Great Plains region of the Southwestern United States. This native tree is being planted more frequently in the home landscape as an alternative to the Australian Pine and the invasive Japanese Yew. The dense, evergreen foliage ranges in color from olive to dark green. A distinctive characteristic of the tree is the variation in its leaf form. The new foliage is prickly and needled shape, and flat scale-like needles occur on the more mature foliage.

Eastern Red Cedar

The female trees produce frosty, gray-green berries that are actually tiny fleshy cones. These provide an excellent winter and early spring food source for a variety of wildlife. Cedar waxwings, robins, bluebirds, turkeys, pheasants and other fruit-eating birds love the berries. Sparrows and cardinals like to build their nests within its dense foliage. It is the host plant for the caterpillar of the Olive Hairstreak butterfly.

The tree is sometimes susceptible to cedar rust, an interesting fungi that forms tufts of bright yellow, jelly-like growths that appear on the tree branches during rainy spells. Do not plant near apple trees, quinces, hawthorns and mountain ashes.

ACTIVITIES AND ADVENTURES FROM
THE BACKYARD AND BEYOND

"Not what you give, but what you share, The gift without the giver is bare…"
– James Russell Lowell

These decorations and ornaments were created from common organic materials gathered with relative ease throughout late fall and winter. Seeds, pods, dried flowers and plant foliage can be found in many shapes, colors and textures.

Oak Gall

Galls are one of nature oddments; they are deformities caused by insects living inside plants. Oak apple and goldenrod galls are two of the most common galls found on plants. Oak apples are formed on the leaves of oak trees in early spring by a family of wasps called *Cynipides*. The small, round, tan galls are easy to spot in the autumn after the leaves have fallen from the trees. The elliptical, goldenrod gall forms on the stem of the plant and is home to a little gray moth caterpillar. They are very noticeable in the winter when the snow is on the ground.

Collecting throughout the four seasons yields a better variety of things to harvest, so don't wait until fall or winter to begin gathering your materials for crafting. Practice stewardship when taking treasures from Mother Nature's cupboard, *"take what you need, but need what you take."*

A tree cookie is a thinly-sliced cross section of a tree branch or trunk. They can be used for a host of fun craft projects. Very thin slices can be made into name tags or place cards for a party. Decorations can be painted, stamped or glued to the cookie's flat surface. The bark on the outside makes a lovely natural frame. Before using, the cookies should be spread out to dry for a few weeks. If you want to speed up the process you can bake them in an oven at 200°F for a few hours. They can get moldy if they are not properly dried. Be sure to cut a few extra to save for other craft projects later in the year.

OAK APPLE SNOWMAN

Materials needed:

- Extra thick tacky glue
- Three brown oak apples
- One small tree cookie, 1-inch or so in diameter
- White acrylic paint
- Two tiny twigs, no larger than 1-inch
- Five black peppercorns
- One kernel of Indian corn
- Three small black beans
- One large burr oak cap
- Dried flowers and berries
- Ribbon

Oak Apple Snowman

Carefully glue the largest oak apple to the tree cookie for a stand. Let it dry. Glue the remaining oak apples on top of the oak apple attached to the tree cookie stand forming a snowman. Let the glue dry for about a half an hour.

Paint the body white and let dry. Cut the tiny twigs to a sharp diagonal point and very carefully push them into the sides of the middle oak apple to form the snowman's arms.

Glue the peppercorns to the snowman's head to make his eyes and mouth; the corn kernel is his nose. Glue a black bean on the front of each oak apple to make his buttons. Glue the burr oak cap on top of the snowman's head for his hat. Decorate the cap with wee bits of dried flowers and berries. Tie a small colorful ribbon around his neck to finish the project.

MILKWEED POD ORNAMENTS

Materials needed:

- Gold paint and watercolors
- Milkweed pod
- Tacky glue

- Assorted pressed flowers and herbs
- Acrylic sealer
- Dental floss and darning needle

Milkweed Ornament

Spray paint the outside of the milkweed pod gold, then paint the inside of the pod a color that will blend with your flowers. Try different ways of arranging the pressed flowers inside of the milkweed pod. After deciding on final design, glue flowers in place.

Seal and finish the pressed flower arrangement inside the pod by spraying it with clear acrylic sealer available at most craft stores. Using the darning needle, thread the dental floss through the top of the pod for a hanger.

TREE COOKIE ORNAMENTS

Materials needed:

Tree Cookie Ornament

- Paint
- Tree cookie
- Drill
- Raffia for hanging
- Tacky glue
- Dried flowers, herbs and assorted seed pods
- Acrylic sealer

You can paint the front of the cookie a color of your choosing or leave them natural. Pre-drill a hole through the cookie at the top to attach the raffia hanger before decorating. Then simply glue an arrangement of dried flowers, herbs and seed pods on the front. The flower arrangement can be elegant or very simple depending on personal preference. Finish and seal the arrangement by spraying with an acrylic sealer.

CHRISTMAS SPIDER

This is a very simple project, which works very well for pre-school children.

The Christmas spider poem, attached to the spider ornament, makes a sweet handmade gift for a child to make and give to family and friends.

Materials needed:

- Extra thick tacky glue

WINTER SOUNDS

Muffled noises
Creaking branches
Rustling leaves
Cracking trees
Lashing wind
Hushed silence
Cawing crows
Whispering pines

- Four pieces of 3-inch brown pipe cleaner
- Two $1^1/_2$-inch corncobs
- Two small acorn caps, $^1/_2$-inch
- Two wiggle eyes
- 12-inch #24 gauge florist wire

Christmas Spider

Glue the two sides of the corncob together to form the spider's body. Bend the brown pipe cleaners in half to make four sets of legs. Dab a bit of glue onto the bent part of each set of legs, then push the glued section of the pipe cleaner into the ends of the corncob's soft pith or inner core. Glue the acorn caps on top of one of the sections of corncob to make the spider's eye sockets. Glue the wiggly eyes into the acorn caps. After the spider is dry, attach a wire around his mid-section for a hanger.

TULIP TREE LEAF ANGEL

Materials needed:
- Two matching golden-colored
  Tulip tree leaves *(Liriodendron tulipifera)*
- One sheet of gold construction paper
- Tacky glue
- One goldenrod gall with stem attached
- Flat toothpick
- Milkweed seed fluff
- Damp cloth
- Bits of pressed flowers, ferns and a ribbon
- Wire hanger used to hang Christmas ornaments
- Two pressed silver lamb's ear leaves

Tulip Tree Angel

Trace the leaves onto the gold construction paper, then cut the paper leaf pattern out about 1/3 of an inch smaller than the tulip tree leaf. Spread tacky glue over the paper leaf and attach it to the back of the tulip tree leaf. The paper leaf provides a sturdy backing for the fragile tree leaf. Let dry.

Glue the stem of the goldenrod gall between the paper leaves for the angel's neck and head. Finish gluing the two leaves together, creating the angel's head and robe. Spread tacky glue on top of the goldenrod gall, using a flat toothpick fasten milkweed seed fluff plugs into the glue on the angel's head. Keep a damp cloth handy to wipe the glue from your hands.

THE CHRISTMAS
SPIDER

A woven silken web,
the child to enfold
Was spun by a spider
in a stable so cold.
In thanks for the warmth
for her shivering Babe,
Between Mary and
the spider, a promise
was made.
Good fortune will follow
all those who can see.
A spider on the
Eve of Christmas
on their Christmas tree.

– ANONYMOUS

Tie a thin brightly colored ribbon around the goldenrod stem at the angel's neck. Attach the Christmas wire hanger to the ribbon at the back of the angel's neck; then attach the lamb's ear leaves to the back of her robe to form her wings. Decorate the front of the angel's robe with the bits of dried flowers and fern. Last, but not least, add one special flower to adorn her soft white hair.

THYMELY TIPS AND SAGE ADVICE

• Before setting your Christmas tree in the stand, cut a disc 1/2- to 1-inch thick off the bottom of the tree trunk. This removes the natural resin that can prevent the tree from taking up water, drying out and dropping needles. Dry the disc (tree cookie). Then date and decorate it to place on your tree as a special family ornament.

• Christmas Gift — something for that special gardening friend, a longed-for plant they would not be extravagant enough to purchase for themselves, delivered at planting time, would make a gift to be remembered.

• Start reading up on wildflowers so you will notice them as they begin to appear in the spring.

• Check over your tools in this slack season, when the spring rush is still far away.

• The purpose of winter mulches is not to keep the cold out but to keep it in and to stabilize the soil temperature by preventing fluctuations. That's why you wait until the ground is frozen an inch or so before putting on the winter covering.

• When you take down your Christmas tree after the holiday, set it up outside and decorate it with popcorn, cranberries, balls of seed and suet, stale bread and apples. Be sure to scatter wild birdseed and bread bits around the bottom of the tree for the juncos and sparrows who like to

WINTER PATTERNS

Snowflakes

Icicles

Ice crystals

Leaf scars

Snow blossoms

Silhouettes

Sharp edges

Brittle edges

dine off of the ground.

• Cut and dry some of the branches of your Christmas tree, then blend the needles with dried, scented geranium leaves, lavender flowers and southernwood to make fragrant sachets to freshen your closets and drawers.

• Cinnamon sticks, coriander seed, allspice berries and orange peels simmered together in a pot on the stove send forth a lovely fragrance to fill the house with Christmas spirit.

• Join in with bird lovers across the United States counting the birds in your backyard and beyond and then record your information at The Great Backyard Bird Count, http://www.audubon.org/Bird/cbc/faq.html.

• Since ancient times fire has been a symbol of home and safety. "Kindling" the Yule log is a holiday custom that originated in Northern Europe with the Druids. An oak log was brought in the house with accompanying festivities on Christmas Eve. Its presence in the house was supposed to ward off evil and the oak ashes were sprinkled on the fields during Christmas week to assure good crops in the coming year. A fragment of the old log was kept and used to light the new Yule log a year later.

As the days lengthen, then the cold begins to strengthen.
– Old Weather Lore

Due to seasonal lag, the earth stores up the cold gathered throughout the shorter days of fall and winter. Air is not warmed just by the sun passing through, but to a greater degree by the ground to which it is near. This is why the overall coldest time of year over the northern hemisphere is not the shortest day of the year but about four to six weeks later.

JANUARY

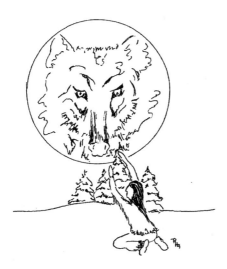

Bare branches of each tree on this chilly January morn
look so cold, so forlorn. Gray skies dip ever so low
left from yesterday's dusting of snow.
Yet in the heart of each tree waiting for each who
wait to see new life, as warm sun and breeze will blow,
like magic, unlock springs sap to flow, buds,
new leaves, the blooms will grow.

– NELDA HARTMANN, JANUARY MORN

Wolf Moon — Very Cold Moon
Cold Makes the Trees Crack Moon

American Indians admired and respected the gray wolf for its cunning intelligence, cooperative spirit and well-ordered lifestyle. When midwinter temperatures dipped below freezing, heavy snows blanketed the ground and the moon shone bright in the frosty sky above, their hungry howls could be heard as they lingered about the outskirts of American Indian villages. Hence, the name Full Wolf Moon. At home in most any habitat, grey wolves were once found throughout the Northern Hemisphere. During the nineteenth and twentieth centuries they were systematically destroyed and almost totally eradicated. Thanks to conservation programs enacted toward the end of the twentieth century they are making a successful comeback across North America.

The names "Very Cold Moon" and "Cold Makes the Trees Crack Moon" aptly describe January, the depth of winter month. Frigid temperatures and deep snow made life difficult for people, animals and plants. Tree branches would crack and break beneath the heavy weight of snow and ice.

Winter was storytelling time in the villages. Elders would gather the young ones around a blazing fire in their wigwams and use stories and songs to connect them with their cultural traditions and ancestors. They believed every bit of rock, each flowing river, each majestic tree, each fragrant flower, each glistening star and each breath of wind was imbued with a spirit. Myths were created to explain the movements of the sun,

Our native landscape is our home, the little world we live in, where we are born and where we play, where we grow up and finally where we are...laid to eternal rest. It speaks of the distant past and carries our life into the tomorrow. To keep this pure and unadulterated is a sacred heritage, a noble task of the highest cultural value.

— Jens Jensen to Camillo Schneider, April 15, 1939

> In the end, we will conserve only what we love: We love only what we understand; and we understand only what we have been taught.
>
> — Baba

the moon, and the stars. Legends explained the origins of plants and animals. "Why-So" stories were used to entertain the children while teaching facts of local natural history, or to illustrate how a specific characteristic of a plant, animal or landscape feature came to be. The power of storytelling was used to share moral and spiritual lessons in a form easily understood by children.

While the stories and myths varied according to geographical region, seasonal cycles and a tribe's way of life, a common cord bound all the tales. Cultivating a thankful spirit, and respect for all living things were the purpose.

Viewed from a distance, a snow-covered landscape appears reticent, lifeless and unfriendly. But to anyone who dares to step outside and check out the wintery world, they will be delighted to discover that adventure lies deeper than the surface appearance. Snow, freezing temperatures and harsh conditions can pose threatening conditions for plants, animals and people. "Dress for success" when going out for a winter hike; proper clothing is an important necessity. Be prepared and wear a warm jacket, hat, mittens and sturdy insulated boots. Layering a sweater under a jacket allows for adjustment depending on exertion and temperature. If you are uncomfortable, shivering and cold you will not be aware of anything else.

Swamps, wetlands and ponds inaccessible during other seasons of the year are frozen solid, providing an excellent opportunity for up-close investigations of the natural features and evidence of wildlife, assuming an adult has checked this out first.

Signs of insects, now concealed in their winter homes, can be observed on decaying logs. Moths and butterflies, warm and cozy, are tucked inside their silken cases or wrapped within hanging leaves attached to plants. They are hiding and awaiting the warmer days of spring. Investigate and

JANUARY
Little January,
Tapped at my
door today
and said, put on your
winter wraps,
and come outdoors
to play.

Little January,
Is always full of fun:
Until the set of sun.

Little January
Will stay a month
with me.
And we will have
such jolly times –
Just come along
and see.

– WINIFRED C. MARSHALL

discover how and where some of the common animals and insects are hiding and spending their winter days.

The nests of birds, squirrels and hornets hidden from view during other seasons of the year are now clearly visible in the dormant trees and shrubs. It is amazing to see all the different materials used to construct them. There are regulations against removing bird nests, so look but leave them where the birds built them. Take photographs to add to your nature journal to remind you to revisit the site during other seasons of the year.

Tracks and trails of animals can easily be observed and followed in a snow covered landscape. Imagine being a CSI agent, follow and identify the tracks, where they lead and then interpret the stories told by them.

People are often taught to identify trees by their leaves in summer and fall. Winter is a good time to identify trees by their silhouettes, twigs, buds, leaf scars and bark texture. *A Guide to Nature in Winter*, by Donald and Lillian Stokes, and *Discover Nature in Winter*, Mechanicsburg, Pennsylvania: Stackpole Books, 1998, by Elizabeth P. Lawlor, are two excellent books with helpful keys to help with winter plant identification.

A host of neat and interesting winter projects can be created from natural materials gathered on a winter hike. Wreaths, arrangements and imaginary creatures can be fashioned from dried flowers, weeds, seed pods and cones. Dish gardens and terrariums can be made with lichens, mosses and various other woodland treasures. Whenever gathering natural materials always be respectful of the site you are gathering from and leave some behind. Practice "green etiquette." Nature always has a reason for everything we see — *"A place for everything and everything in its place."*

Winter Weed Walk and Scavenger Hunt

A winter weed walk will reveal much about plant structure, form and habitat. To get the most from hunting and observing winter weeds it is important to be aware of all their points of interest and natural history descriptions. Begin with the common features and characteristics that are readily observable.

People of all ages enjoy scavenger hunts. Try adapting the information and points of interest listed below into a scavenger hunt. Design the list so that it requires people to take an up-close and personal look at the diversity, colors and shapes of common winter weeds in your backyard and beyond.

Winter Weeds

Tough and resilient, weeds can be found in a vast variety of habitats — woodlands, roadsides, the edges of parking lots and even in the cracks of sidewalks. Many of the plants we classify as weeds were deliberately brought into the United States because of their herbal uses. Escaping from cultivated gardens they have become naturalized across North America. Others are native plants that have existed in an area since before European settlement. Every plant has a story to tell and when viewed together in the landscape they reveal the regional history of how the land was used by past generations of people and wildlife.

Points of Interest:

- Are the stems smooth, prickly, hairy or velvety?
- How are the stems shaped? Square, triangle or round?
- Look for unusual or remarkable characteristics a plant might possess such as thorns or burrs, translucent seed pods, opposite branching,

Winter is thought of as a pause, a time of retreat and waiting. The seeds that fell last August are supposedly dormant. Yet not altogether dormant, these too are acted on by the season. Many seeds need this period of cold before they can sprout… On the chilly ground, under the snow perhaps, the seeds are alive and in the act of becoming; the thrust of the cold and dark are as necessary to the scheme as light and warmth. In the rough circles of seed shapes life is staked out; the property of life itself is claimed.

— Anita Nygaard,
Earth Clock, 1975

branching only at the top or no branches at all.

• Can you see evidence of animals or insects interacting with the plants? Tracks in the snow around the plant?

• Can you find insect galls or cocoons wrapped inside leaves or egg cases attached to plant stems or stalks?

• When you crush the plant, is there a notable fragrance?

• Bring along some paper envelopes for collecting weed seeds from a variety of sites. Identify the seeds and then label the envelopes with the seed name and location found.

• Seeds that float with plumes or down — milkweed, dandelion, cottonwood, thistle.

• Seeds that fly with wings — maple, tulip, popular, ash.

• Seeds that cling or are sticky — goldenrod, burdock, cocklebur, foxtail.

• Seeds that animals like to eat — cherry, crabapple, pine.

• Exploding seeds — witch hazel, lupine, jewelweed.

What human inventions might have been inspired by some of these

seeds and their dispersal methods? Plant some of the seeds and see what happens.

Plant proverbs and gardening maxims rooted in common sense or practical experience conveying general truths or moral lessons have been passed down through the ages.

"He that goes barefoot must not plant thorns"; "The pine wishes herself a shrub when the axe is at her root"; "A weed that runs to seed is a seven years weed"; "Oaks may fall when seeds brave the storm" and *"Make hay while the sun shines"* are a few samples of practical words of wisdom. Others, perfectly understandable in times past now seem meaningless, *"He that would live for aye, Must eat sage in May."*

Make a collection of plant proverbs and seed sayings and compare the connections between the plant's growth habits and human behavior. Write down some plant proverbs or maxims that can be applied to human behavior using popular plants in the twenty-first century.

The January Thaw

Everyone loves to talk about the weather. We are all familiar with seasonal tales predicting annual spells of hot and cold weather. Seasonal folklore special to North America forecasts an annually occurring warm spell sometime during the third week of January. Every generation since weather tales have been told has believed or experienced this January thaw. Historical records and journals from past generations appear to confirm this regularly recurring weather pattern in the northern hemisphere. A welcome break from the raw, cold, dark days of winter, reminding us that spring is just around the corner. An online article entitled the *Weather Doctor Almanac, 2002* states:

> The January thaw, according to the 1954 Glossary of Meteorology is: A period of mild weather, popularly supposed to recur each year in late January in New England and other parts of the Northeastern United States… statistical tests show a high probability that it is a real singularity.

Other recurring examples of folklore seasons are the Dog Days of Summer, Indian Summer and Blackberry Winter.

The Weather Doctor, http://www.islandnet.com/~see/weather/doctor/htm, is a great Web site. Be sure to check it out for more fun weather facts and strange tales for every month of the year.

Mosses and lichens are members of the division of flowerless plants that reproduce by spores, a huge contrast to the complex structure of flowering plants. These primitive plants often go unnoticed, because they do not seem as glamorous as the flowering plants that adorn our gardens

All winter long
behind every thunder,
guess what we heard!
behind every thunder,
the song of a bird
a trumpeting bird.

All winter long beneath
every snowing
guess what we saw!
beneath every snowing
a thaw and a growing,
a greening and
a growing.

Where did we run
beyond gate and
guardsman?

Guess if you can
All winter long
We ran to the sun
the dance of the sun!

– NATIVE AMERICAN SONG

and landscapes throughout spring, summer and fall. A January thaw is a golden opportunity to introduce children to these tiny treasures which are often dormant in other seasons of the year.

Moss and lichen can be found happily growing together whenever temperatures rise above freezing. They grow in poor soil, on the edges of forests, on rocks, old stumps and trees. Mosses and lichens are thought to be some of the first plants to live on the land. They are often called a pioneer species because they are frequently the first plants to begin colonizing newly exposed rocks and disturbed soils. They hold moisture, add organic matter and softly cover uninviting barren spots until new and more diverse plant species can gain a foothold. They are tough little plants and can survive in very harsh environments, but they do have their limits. They are very sensitive to air quality. Scientists consider lichens and mosses valuable aids for monitoring air pollution. Patches of gray lichens and green moss growing on older trees can be an indicator of good air quality. Tardigrada, plump soft-bodied microscopic critters commonly called "moss piglets" or "water bears" live in mosses and lichens. They can easily be seen padding about their business with a microscope. To learn more about these cute cuddly creatures visit http://www.microscopy.uk.org.uk/mag/artmay99/dwbear.html.

Mosses belong to a group of green plants called Bryophytes; they have no vascular systems or flowers. Even though mosses do not have flowers, their fruiting stems release spores that are very interesting to look at if you have a hand lens. You can easily find two basic types of moss, those that form dense pads and those that resemble soft green, spreading carpet. The hair-cap or pigeon wheat is a carpet moss that can be found growing in all parts of North America. Children used to refer to this plant as fairy forests and played with the fruiting stem of this moss to make ephemeral rings and necklaces.

Mosses

A mushroom went into a coffee bar and saw some algae around a table. He went up to one and said, "you're a good looking all gal (algal)." She looked him over and said "you look like a fun guy (fungi)," and they took a liken (lichen) to each other.

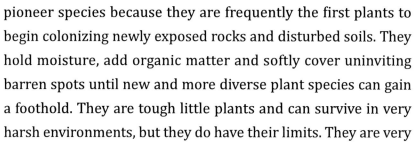

An interesting video about the life cycle of a moss can be viewed at http://www.sumanasinc.com/webcontent/animations/content/moss.html.

Lichens are a combination of two plants that have formed a symbiotic relationship. It has been described rather like a pie: the fungus forms the crust and the alga is the filling. They have a soft natural fragrance and are used in the manufacturing of perfumes. Hummingbirds use them to build their nests and American Indians would wash their babies in lichen-scented water. Lichens display visual proof there is "Strength in Diversity."

On a winter walk you can find tiny seedlings of evergreen trees, winter rosettes of biennial weeds, pads of moss, pieces of lichen encrusted rocks and sticks, all great to collect and use for an indoor winter gardening project.

Terrariums and Dish Gardens

Terrariums and dish gardens are fun to plant in any season, but January is an especially fine time to plant them with materials gathered during a January thaw. A terrarium is a small, closed ecosystem growing in a transparent container with a tightly-fitting removable lid. Glass aquariums, goldfish bowls, old condiment jars or two-liter pop bottles can all be used for terrariums. Plants will live indefinitely in a stable, unchanging atmosphere, provided the humidity and temperature does not vary greatly over time. Dish gardens do not have covers and are a simple form of miniature landscaping.

When planting a woodland terrarium or dish garden, it's very important to choose plants that like similar growing conditions. Growing conditions include light, humidity, temperature, water and soil requirements.

Before you start, organize all of your materials and plants in one place. It's always a wise idea to experiment with your plants in different landscape designs before finally planting them in your container.

TERRARIUM

Materials needed for container and plant material:
- Container with lid
- Pea gravel or very coarse sand and activated charcoal
- Sphagnum moss
- Moist potting soil
- Rock
- Fork and spoon
- Plants
- Plant mister

1. Cover the bottom of your container with an inch of coarse sand or pea gravel, with a half-inch of charcoal on top of the gravel to help filter the air. A thin layer of sphagnum moss placed over the gravel and charcoal will prevent the soil from sifting into the drainage space.

2. After layering the drainage materials evenly across the bottom of the container, add two or three inches of potting soil at different levels to create terraces, valleys and hills. Then add in some of the larger decorative pieces of rock and natural materials.

3. Use a small kitchen spoon and fork for a shovel and rake to help plant your plants in the soil, being careful to allow sufficient space for them to grow and spread. After planting, mist the plants with water to wash off any soil from their leaves and sides of the container. No heavy watering is needed. Let set for a few hours and then close the container. Occasionally seedlings will appear in the container that were in the soil collected with the plant material. Leave them and watch to see how they adjust to the environment in the container. Compare them with the seedlings that emerge in the spring in the location they were originally found.

4. Terrariums should never be grown in direct sunlight and grow well by windows facing west or northeast. If too much condensation forms inside the container, remove the lid for a few hours.

Dish Garden

DISH GARDEN

You can use almost any shallow dish you have around the house. Cover the bottom of your dish with a thin layer of pea gravel and then add a couple of inches of moistened soil over the gravel, use your fork to level the soil in the dish. Arrange the moss and lichens in an interesting design, use little rocks and twigs to build fences and pathways. Add tiny animals and birds to create the illusion of a living landscape. Tuck some grass seed in a few spots and after it germinates keep it clipped at different levels to add height and depth to your mini-landscape. Water as needed and grow in indirect sunlight.

GOOD FRIENDS — OUR NATIVE TREES

Eastern White Pine *(Pinus strobus)*

Tall, stately and soft-spoken, the white pine can trace its family lineage back through time to the Devonian age, a distinguished characteristic shared by fellow friends, mosses and lichens. White pines are the tallest of the northeastern trees. Before pre-colonial settlement, vast tracks of huge majestic white pines covered much of eastern North America. Historical records dating back to the 1700s describe towering stands of white pine trees over 250 feet tall. When colonists arrived in America from Great Britain, the white pine became their wood of choice for making masts and carving figureheads for their sailing ships. American Indians built thirty to forty-foot canoes from pine logs, waterproofing them with pine resin mixed with beeswax. They collected the inner bark during the winter and made a type of flour to help stretch their food supplies. Shaved pine knots were used to make a cure for poison ivy.

The white pine is wind pollinated and has soft, slim bluish-green needles, two- to five-inches long in bundles of five. It produces slender three- to four-inch cones that take two years to mature. Poetry and folklore abound with references alluding to murmuring, whispering or sighing sounds associated with pine groves. The shape and arrangement of the

Many voices there are in nature's choir and were good to hear, Had we mastered the laws of their music well, and could read their meaning clear;

But, we who can feel at nature's touch, cannot think as yet, with her thought; And I only know that the sough of the pines with a spell of its own is fraught.
— FRASER'S MAGAZINE

Eastern White Pine

needles cause this phenomenon, *"The needles of the pine act like the strings of an Aeolian harp; and the wind, in passing through the tree, sets them into vibration, making a sighing sound which seems to the listener like the voice of the tree."* (Comstock, 1955 p. 672)

White pine needles contain high levels of vitamin C and were used by American Indians to make a winter cold remedy. In the spring and fall they would cleanse their homes with pine needle smoke to prevent sickness. Sweet smelling boughs of pine were used for mats, rugs and bedding and Native American children would fashion dolls from the needles.

Eastern white pine is one of the fastest growing conifers in the upper Midwest. It is widely cultivated on tree farms across Eastern North America. Because of its evergreen nature, the white pine is a symbol of the unchanging aspects of life and is widely planted by landscape designers to add a tall spark of green to the winter landscape. It is also a good tree to plant if you want to attract wildlife to your yard or neighborhood. It is an important food source for birds such as chickadees, grosbeaks and nuthatches. They feed on the seeds, as do chipmunks, red and gray squirrels, turkeys and the meadow vole. The bark and needles provide a food source for rabbits, beavers, porcupines and deer.

Black bear cubs climb white pine trees for safety when they feel threatened. The Eastern Pine Elfin butterfly lays her eggs on the new needles of young white and jack pine trees. The caterpillar is dark green with two cream-colored stripes mimicking the color of the pine needles they feed on after hatching.

ACTIVITIES AND ADVENTURES FROM THE BACKYARD AND BEYOND

• Select a site in your yard that will make a good palette for creating a work of ephemeral art that can also be viewed from an inside window. Gather twigs, sticks, snow, leaves, dried foliage, pine cones and pods or whatever catches your eye. Assemble your materials in one spot and begin creating your work of art. Keep in mind that ephemeral works of art are short lived and transitory in nature, part of their beauty is watching them change from day to day. Mother Nature is a master at ephemeral works of art, she has a special show for every season. Just follow her around. To learn about ephemeral art and artists check out this Web site, http://www.morning-earth.org/HowLearn.html.

• Winter is a good time of the year to make tree bark rubbings. All you need is lightweight paper and crayons with their paper removed. Trees with smooth bark tend to produce the best rubbings. Hold the sheet over the bark and rub the side of the crayon over the bark of the tree and a pattern will appear.

• Make a name plaque using your bark rubbings, twigs and white pine needles. Cut a piece of cardboard in a shape you like. Cover the cardboard with one of your lighter-colored bark rubbings. Glue it down and let it dry. While your plaque is drying, experiment how to spell your name with the twigs and pine needles. Then glue them on top of the bark rubbing sheet.

• Have you ever noticed your dog or cat standing still in the snow, with their head cocked to one side as if listening to a voice speaking from another world? They are probably listening for the sound of a meadow vole running through its tunnel beneath the snow. Meadow voles spend much of their winter creating a labyrinth of snow tunnels. They provide a safe environment for the voles to spend the cold winter months. When the temperatures warm, these tunnels are easy to see and track. Tracking the tunnels may reveal where they have been sleeping and what they have been eating and how they escape from predators who might be waiting and listening to catch them.

• Create a nature journal and begin recording the natural world as it unfolds daily around your yard and neighborhood. Pick out a plant or place that you can follow as it changes with the seasons. Be sure the plant or place is in a location where you frequently spend time. Start now observing and documenting the events that take place involving the plant or place. Take pictures and make a photo essay of the changes that happen from one season to the next. If you don't have a camera, draw or sketch your observations. Include written notes of your impressions and feelings about the events you observe.

JANUARY
So the first month of the year, like its namesake, looks back over the past, and forward to the future, with hope and resolution
– ANONYMOUS

THYMELY TIPS AND SAGE ADVICE

• Take time to sit down as a family and consider just what you would like your home landscape and gardens to give you, and what you would like to give to your landscape and gardens. Develop a four season strategy to ensure your landscape steadily progresses from one season to the next, reflecting the colors, forms and relationships that naturally take place as nature moves from one season to the next. Think about what you need and what nature needs.

• Create a bloom sequence chart of plants currently growing in your yard and gardens. Keep in mind ornamental features and appeal to wildlife vary with plant species; some are attractive to humans or wildlife during just one season of the year while others have distinctly different ornamental characteristics or value in all four seasons.

• *Bringing Nature Home, How You Can Sustain Wildlife With Native Plants*, by Douglas W. Tallamy, is an excellent resource guide for the home gardener looking for information on how to use native plants.

• January is National Mail-Order Gardening Month.

• Vegetables that belong to the same family can be susceptible to the same diseases and pests. Plan in advance how you will rotate your plants from one year to the next. This will get you off to a good start when the planting season arrives.

• Keep an eye out for fungus gnats and scale insects on your indoor plants. Mix up this simple solution to rid your plants of scale and white-flies: one gallon distilled water, one-half cup isopropyl alcohol, one cup ammonia, one teaspoon liquid dish soap. Mix everything together and use a spray bottle to apply to infected plant material.

• Two fun Web sites to explore historical weather, holiday traditions and gardening tips are: http://www.thealmanack.com and www.farmer-salmanac.com.

FEBRUARY

Still lie the sheltering snows, undimmed and white;

And reigns the winter's pregnant silence still;

No sign of spring, save that the catkins fill,

And willow stems grow daily red and bright.

These are days when ancients held a rite.

Of expiation for the old year's ill,

And prayer to purify the new year's will.

— HELEN HUNT JACKSON,
A CALENDAR OF SONNET'S: FEBRUARY

Ground Squirrels Run Moon — Hunger Moon — Opening Buds Moon

The February sunshine
steeps your boughs,
and tints the buds and
swells the leaves within.
— WILLIAM CULLEN BRYANT

The variety of descriptive February full moon names reveals how closely American Indians observed the climate and ecology accompanying seasonal change in their particular regions. February is the last of the three winter months and one of the most miserable and unpredictable in the cycle of the seasons. Winter cache pits of dried corn, beans, sunflower seeds, squash and meat stored during the previous spring and summer have now begun to run dangerously low. Harsh weather conditions and deep snow in some of their areas made hunting difficult, and so by February many of the villages lost members due to cold or hunger; thus the name Full Hunger Moon.

Animals have different ways of surviving in cold winter weather. Many migrate to warmer climates, and others adapt to cold temperatures by remaining active all year long. Some go into hibernation, their bodies shutting down for the winter and waking up with the arrival of spring. The ground squirrel is one of the mammals known as "The Seven Sleepers", animals who go into a deep sleep or hibernation to escape the winter cold. The groundhog is the most famous of the seven, with the yearly anticipation of spring dependent on him seeing his shadow. The little brown bat, bear, ground squirrel, jumping mouse, raccoon and skunk are also listed among the seven sleepers.

Squirrels are divided into three main groups — tree squirrels, ground squirrels and flying squirrels. Ground squirrels include chipmunks, prairie dogs and marmots. Ground squirrels rarely climb trees, and they nest in burrows in the ground or beneath rocks or logs. Squirrels are lively little fellows, but are most active in late winter or early spring when the mating season begins. Resourceful and prudent, ground squirrels also prepare for winter by putting aside a winter cache of seeds and nuts in their winter burrows. Squirrels only find about ten percent of the nuts they hide for safe keeping. Because of their industrious but forgetful nature there is always a new generation of trees and plants springing forth each spring. The Lenni Lenape clans living along the Delaware River named their February moon "When The Ground Squirrel Runs."

All types of trees have different bud patterns and leaf scars that can be used for tree identification in the winter. You can tell the difference between leaf buds and flower buds by their shapes — flower buds are often round and short and leaf buds are long and pointed. Buds mark the end of one season's growth and contain the template for next season's leaves, stems and flowers. Dormant buds produced during the previous summer were closely observed for signs of change. When they began to swell and open in response to the longer daylight hours, hope would also begin to swell among the tribes because the Opening Buds Moon signaled the end of winter and warmer days ahead.

"Project Bud Burst" is an interactive citizen science program with activities for all ages. Check it for great tips on how to read your local landscape, http://www.budburst.ucar.edu.

Henry Ward Beecher once told a story of a little bud who thought she could not unfold when springtime came: And, the sun and the wind laughed, for they knew that when they should shine and blow upon the bud and fill up and swell those tiny leaves, it would open from the necessity of its nature.

Leaf Bud

Backyard Nature Trail

We can all help contribute to a healthier environment one backyard at a time. Traditionally, people have considered their landscapes and gardens separate from the natural environment. A positive shift in thought has happened over the past decade. We are now creating landscapes that blend with and support the natural environment. Nature friendly landscapes are people friendly landscapes, welcoming self-discovery, optimism and a sense of wonder about the natural world.

Building a self-guided nature trail in your backyard is an exciting open-ended project that will inspire you to view your yard and neighborhood

A FABLE

The mountain and
the squirrel
Had a quarrel,
And the former called
the latter, "little prig":
Bun replied,
You are doubtless very
big, But all sorts of
things and weather
Must be taken in
together, make up a
year, And a sphere.
And I think it no disgrace
To occupy my place.
If I'm not so large as you,
You are not so small as I,
And not half so spry:
I'll not deny you make
A very pretty squirrel
track; Talents differ;
all is well and wisely put;
If I cannot carry forests
on my back,
Neither can you
crack a nut.

– RALPH WALDO EMERSON

in a new way. Working on the trail throughout the year is a wonderful way to explore the different micro-climates within your yard while closely observing the changing plants and animals as each season progresses to the next.

Start by taking a walk around your yard and inventory the special features already in place — trees, both dead and alive, rocks, wooded sections and wetland areas. Then look for sites that have potential for habitat enrichment, such as transitional zones between neighboring property lines and public green spaces.

Draw a detailed diagram of your yard listing the trees and shrubs and interesting features. The diagram will show you how plantings can be developed and how the nature trail can make use of the plantings. A winding trail creates a sense of mystery, solitude and spaciousness, so as you lay out your trail avoid straight lines. If it is not possible to lay down an actual trail, use markers, signs or a map to guide people along the trail. Elevated, shady and hidden spots make good observation points to place a bench or old log for people to sit, listen and connect to their surroundings.

More wildlife is always found where two habitats meet and blend than at the center of either one. All wildlife has four basic requirements for survival: food, shelter, water and safe places to raise their young. Combinations of these four elements must be thoughtfully planned for each species of wildlife you are trying to attract. Another important consideration is selecting plants that overlap in flowering and fruiting times so there is a good food supply available throughout the year.

Visit the National Wildlife Federation Web site for more help and detailed information, http://www.nwf.org/In-Your-Backyard.

Spiders, caterpillars, beetles, ladybugs and crickets prefer building their homes and raising their families in the tall grass along the edges of gardens and yards.

Birds, butterflies and small mammals need trees and shrubs for cover, food, nesting and raising their young. Leave some stands of brush, piles of leaves, rocks and stacked logs for small mammals, insects and birds to hide and watch the people walking along the trail

After your trail is laid out, a few well-placed interpretive signs will help focus attention on the special natural features or wildlife habitats you have identified along your trail. Briefly note on the sign, depending on the season, the worthy things people should be looking for. Posts or dead trees drilled with carefully located peepholes will invite observers to take an up-close look at specifically selected items or views you would like them to see.

You know your street address and city you live in but do you know your Ecological Address? Check out this Web site for interesting natural components that make your home and surrounding neighborhood here on the earth unique: www.audubon.org/bird/at_home/Explore.htlm.

Building a backyard or neighborhood nature trail with family and friends is an excellent and rewarding way to become familiar with the common plants and animals sharing your backyards and neighborhoods. The trail will also provide safe passages for migrating wildlife to stop, rest and eat while traveling.

Let Nature be Your Guide

Groundhog Day is a North American folk tradition that settlers from Germany and Great Britain brought to America in the 1800s. Early American farmers believed it was a day to forecast the next six weeks of weather because the day marks the midpoint of winter, halfway between the shortest day of the year and the spring equinox. It was the day to inventory their winter supplies; if they did not have *"half the wood and half*

> Whether people are fully conscious of this or not, they actually derive countenance and sustenance from the 'atmosphere' of the things they live in or with. They are rooted in them just as a plant is in the soil in which it is planted.
>
> – Frank Lloyd Wright

the hay" left, tough times might be ahead before spring and fresh green grass was available. According to old weather lore, the groundhog awakens from its long winter sleep on February 2. He pokes his head out of his underground burrow, takes a good look around and tests the weather conditions. If the sun is shining and the groundhog can see his shadow, he gets scared and drops back down into his burrow. This supposedly means there will be six more weeks of winter. If the day is cloudy and overcast and he can't see his shadow, he leaves his burrow, indicating that spring weather will soon come. This prediction owes its origin to the European tradition of Candlemas, a Christian festival honoring the purification of the Virgin Mary after the birth of Jesus. For centuries it was the custom to have the clergy bless candles and distribute them to the people, symbolizing the words of Simeon, to Mary, that Jesus would be a *"light to the gentiles and the glory of the people of Israel."* Scriptural References: Luke 2:22-39.

An English song goes:

If Candlemas be fair and bright,
Come winter, have another flight;
If Candlemas bring clouds and rain,
Go winter, and come not again.

Before there were clocks, ancient people used shadows to tell the time of day and year. They first learned to measure time by the division between day and night. They noticed the sun always came up over the eastern horizon and set each evening in the western horizon, covering the world in darkness. During the day they observed that a shadow cast by a tree or rock was longer early in the morning and grew steadily shorter until it disappeared. When the sun was directly overhead at midday, they also noticed that the shadow grew longer again, on the other side of the tree or rock as nightfall drew closer. They knew the shadows cast by the sun changed from day to day and month to month, so they knew the time of year by noting the location of the sun as it rose each day.

Using the Sun and Shadows

The first instrument used to measure time was a stick placed into the ground with marks in the dirt to show where the stick's shadow was every hour. You can use shadows to determine both direction and the time of day. A shadow stick is a primitive form of a sundial.

On a sunny day, place a stick upright in a bare spot of ground. Mark the end of the stick's shadow with a rock. The first shadow mark is always west. To determine a true east-west line on a sunny day, place a stick upright in a bare spot of ground. Mark the end of the shadow tip with a stone.

Wait 20 minutes or so, then again mark the end of the shadow tip again with another stone and draw a straight line between the stones. This line runs roughly east to west.

Place your left heel on the second stone, with your toes pointing towards the first stone. Your foot should be pointing due west now and if you put your right arm straight out from your side, it will be pointing due North.

As midday approaches, the shadow will shrink and disappear. Then in the afternoon it will begin to lengthen again.

Long before clocks became common, many rural homes had a noon mark near some southern window or door of their homes, telling members of the house when midday (solar noon) had arrived. This noon mark alerted everyone in the home when their day was half over and they had a set amount of daylight left to finish their tasks. A noon mark is a very simple device that can be put on any wall that catches the sunlight at midday at its southern most spot in its daily transit across the sky. But,

Many, many welcomes,
February fair-maid
Ever as old time
Solitary firstling,
Coming in cold time
Prophet of the gay time
Prophet of the May time
Prophet of the roses
Many, many welcomes
February fair-maid.

– Tennyson

only the time at noon is shown. There are supposedly four times a year when sun time and clock time agree: April 15, June 15, September 1 and December 24. For more information on how to make a noon mark or sundial visit http://www.sundial.co.uk/home3.htm.

Finding Your Way Without Map or Compass, New York: Dover Press, 1983, by Harold Gatty, is an immensely instructive book for anyone who enjoys exploring the outdoors. It is filled with fascinating examples of how primitive cultures used the sun's angle, the night sky, wind direction and plants as navigational guides for measuring time and travel.

GOOD FRIENDS — OUR NATIVE TREES

Pussy Willow *(Salix discolor)*

Willows are a large family of trees and shrubs that grow along streams and other moist locations throughout the United States. The appearance of willow catkins is a welcome sign of spring. A catkin is a dense flower cluster typified by the pussy willow. You have to look closely to see the small, simple flowers that are often greenish-yellow in color and appear before the leaves do. The willow is the exception to the rule that only plants with showy complex flowers have their pollen distributed by insects. Catkins can also be found in early spring drooping from branches of birches, beeches, hazelnuts, hornbeams, mulberries, chestnuts and even some oaks. These trees rely on the wind to pollinate their primitive flowers because they have nothing to offer as tasty as willow nectar.

Pussy Willow Bouquet

Willow bark contains the active compound salicin, used in many folk medicines. Aspirin is a derivative of salicylic acid which was first synthesized from derivatives of willow bark. The willow is often given the nickname "toothache tree" because American Indians chewed willow bark to help reduce pain and fever. They wove baskets from long straight willow branches. The bark was used to make rope, fishing nets and ties for fastening . Children made whistles from the branches in early spring. The stems were used to make hunting bows, arrows and to roast fish. Forked sticks made of willow were sometimes employed in the ancient art of water witching. Today the light hardwood is used to make wooden limbs and cricket bats. There are many tales and superstitions associated with willow trees. In the language of flowers, willow branches symbolize mourning and forsaken love. The following folk tale is one of many passed down through the ages explaining the origin of pussy willow catkins.

Legend of the Pussy Willow

According to an old Polish legend, many springs ago there were willow trees, but they did not have any catkins. A farmer, annoyed that his cat had yet another litter of kittens, decided to toss them into a flowing river not far from his farm. The mother cat followed him and sat crying on the side of the beautiful flowing river edged with willow trees. As the farmer prepared to toss the kittens into the flowing water, the willows heard the mother cat's anguished cries. JUST as the farmer opened the bag and began to drop the kittens in the water, the willows dropped down their long branches into the water. The kittens held onto the branches and scrambled safely up the banks beside the flowing river. Every spring since then, as the legend goes, the willow branches sprout tiny fur-like buds along the branches where the kittens once clung.

Willows are easy-going plants that come in all shapes and sizes. You can find many varieties with attractive stems and interesting catkins. Two species of willow which are commonly called pussy willows are *Salix discolor*, native to North America, and *Salix capera*, native to Europe. Both can be forced in the early spring and then dried for later use in

Then saffern swarms swing off from all the willers. So plump they look like yaller caterpillars.
– JAMES RUSSELL LOWELL

crafts and flower arrangements. The black willow, *Salix nigra*, is a common native willow with a large crooked leaning trunk and graceful branches and is often seen growing along water routes across North America. Mature willows are notorious for having weak wood that breaks easily with ice, wind or age, thus the old adage *"Willows are weak, yet they bind other wood."* Pussy willows have an untidy growth habit and work well as an edge planting or massed together in damp, out of the way sites. Willows provide a valuable food source of pollen and scented nectar for early emerging insects. Solitary bees, bumblebees and Mourning Cloak butterflies can be found around willows in early spring.

ACTIVITIES AND ADVENTURES FROM
THE BACKYARD AND BEYOND

Bird Feeder With
Suet Mixture

FEEDING STICK FOR THE BIRDS

A simple eco-friendly birdfeeder can easily be made from a recycled piece of tree trunk or branch.

Materials needed:
- A 12-inch length of wire and a marking pen
- A stout section of a tree branch 4- to 8-inches thick cut into a $1^1/_2$- to 2-foot length.
- Wire and a drill with $1^1/_2$-inch bit and a smaller half-inch bit.
- A suet and seed mixture or crunchy peanut butter

Mark one-inch circles all around the log, a few inches apart at frequent intervals. Using the $1^1/_2$-inch drill bit, drill holes into each circle $1^1/_2$-inch. Drill a smaller hole through the top of the stick. Stuff the holes with a seed and suet mixture or crunchy peanut butter. Thread the wire through the small holes at the top of the feeder. Hang outside in a sheltered spot for the birds. This makes a great group project that doesn't cost much but is very attractive hanging in a yard or garden.

February is a good time to clean out your birdhouses and plan locations for new birdhouses. Be sure to look at the different plans available for each species of bird to help you decide which birds you would like to welcome as tenants.

SPRING FLOWERS IN FEBRUARY

Branches of some spring flowering trees and shrubs will bloom now if you cut them and bring them indoors.

Select and cut heavily-budded branches from around the sides of shrubs that are roughly a half-inch in diameter. Place them in a deep container of water and set them in a cool room until the buds begin to swell.

After the buds begin to swell, bring them out into the light and watch them break into blossom. Forsythia, alder, spice bush, blueberry, flowering quince, pussy willows and red maple will all respond to this "forcing" treatment.

Forestfarm is one of the finest plant nurseries online, offering ornamental and useful plants from around the world, and several specialties including native plants and plants for birds and wildlife. Their catalog not only contains useful plant descriptions and cultural information, but also includes recipes, gardening tips and quotes that make it especially fun and inspiring to browse. The catalog is published in the spring and the fall. To add your name to their mailing list go to www.forestfarm.com.

This simple paper-making recipe is from the 2010 Forestfarm Spring catalog.

MAKING PAPER YOURSELF

Paper, like so many things in our lives, is made from plants. There are many possible ways to make paper. One easy way is:

Mix 1 tablespoon liquid starch with one-third quart water in a quart jar. Shake.

Tear 14 sheets of toilet paper (made from plants!) into small pieces and add them to the jar. Shake until a slurry is formed. You can now add bits of lichen, bark, etc., if you'd like.

You can also use up to 50% recycled paper by tearing up old paper (colored paper is fine; in fact, it will add color to your own) into small pieces. Fill a blender half full with the paper, then fill the blender with water and blend it into a slurry. Add to jar above.

Take two plastic sheets of needle point mesh screen and two embroidery hoops (or other frames), slightly smaller than the screen. Place one hoop in a tub or bucket; put a screen on top of the hoop, topped by another hoop. Pour some slurry evenly into dried leaves (Japanese maple leaves would be pretty!) or flowers on the screen, and pour just a bit more slurry over them. Pick up the hoops and screen combination and let them drip into the tub/bucket for a while.

Remove the hoops and put the second mesh screen on top of the screen holding the paper slurry on it. Place the screens, with the paper-in-progress sandwiched between them, on a piece of folded newspaper. Pat with a sponge to remove excess water.

Now place the two screens on a piece of felt and cover them with another piece of felt. Roll gently with a rolling pin to remove yet more water.

Very carefully remove the new paper (it's fragile until it dries) from the screens and place it on newspaper, or something absorbent to dry.

CONGRATULATIONS! You have made homemade paper!

THYMELY TIPS AND SAGE ADVICE

• Herbs are delightful plants that are used for flavoring, fragrance, medicine, decorative crafts, dyes and beauty products. Herbs are not vegetables, grains or lumber-producing trees. Herbs should be included in every home landscape. The Herb Society of America is an educational organization concerned with the cultivation of herbs, the study of their history and their uses both past and present. Visit their Web site at www.herbsociety.org for information and resources for growing herbs in your particular region of the country.

• The choice of plants and their varieties available in seed far outshine the plant selections found at most garden centers. Annual flower and

If we want children to flourish, to become truly empowered, then let us allow them to love the earth before we ask them to save it.

– DAVID SOBEL,
HTTP://ARTS.ENVIROLINK.ORG/
ARTS_AND_EDUCATION/DAVID-
SOBEL1.HTML

vegetable seeds make a good choice for the first-time gardener because they mature quickly and are easy to tend. In addition to saving money, many of our annual flowers, vegetables and herbs perform their best when grown from seed sown directly into the garden. The Home Garden Seed Association (HGSA) has created a list of the top ten easiest plants to grow from seed directly in the garden. For more information check out their helpful Web site at http://www.ezfromseed.org.

• Founded in 1920, The National Garden Bureau offers valuable gardening advice and fact sheets for gardeners interested in growing plants from seed. They also have good information and resources for people working with youth gardening programs. Visit their Web site at http://www.ngb.org/gardening/fact_sheets.

• Climatic change and rising populations are straining water resources everywhere, so state and local restrictions on water use are becoming increasingly commonplace. Landscaping with native plants helps reduce water usage. Check out the Grow Native Web site at www.grownative.org to learn more about native plants and the positive and productive ways they contribute to our local environments.

• A recent NASA experiment confirmed the important role indoor plants play in helping to clean indoor air. House plants removed 87% of toxic indoor air in a closed chamber. Www.o2foryou.org is a very informative Web site with information on how plants help clean our air both indoors and out.

Wet February
next comes by
With chill, damp earth,
and drippy sky;
But, heart cheer up!
The days speed on.
Winds blow, sun shines,
and thaws are gone.
And in the garden
may be seen.
Up-spring flowers
and budding green.
– W. Howitt

Suggestions for Further Reading

Blanchan, Neltje. *Natures Garden.* New York: Doubleday, Page & Company, 1907.

Bruchac, Joseph. *Native Plant Stories.* Golden, Colorado: Fulcrum Publishing, 1995.

Caduto, Michael J. and Bruchac, Joseph. *Keepers of the Night: Native American Stories and Nocturnal Activities for Children.* Golden, Colorado: Fulcrum Publishing, 1994.

Chesanow, Jeanne R. *Honeysuckle Sipping: The Plant Lore of Childhood.* Camden, Maine: Down East Books, 1987.

Cloyd, Raymond A. and Nixon Philip L. and Pataky Nancy R. *IPM for Gardeners: A Guide to Integrated Pest Management.* Portland, Oregon: Timber Press, 2004.

Dowden, Ophelia Anne. *Wild Green Things In The City: a Book of Weeds.* New York: Thomas Y. Crowell Company, 1972.

Dubin, Lois Sherr. *North American Indian Jewelry and Adornment: From Prehistory to the Present.* New York: Harry N. Abrams Inc., 1999.

Dunn, Fannie W. and Troxell, Eleanor. *Mother Nature Series: By The Roadside.* New York: Row, Peterson and Company, 1928.

Dyer, T. F. Thiselton. *The Folk-Lore of Plants.* (First published in 1888) Charleston, North Carolina: Biblio Bazaar Reproduction Series, www.bibliobazaar.com.

Earle, Alice Morse. *Sun Dials and Roses of Yesterday.* New York: The Macmillian Company, 1902.

Erdoes, Richard and Ortiz, Alfonso. *American Indian Myths and Legends.* New York: Pantheon Books, 1984.

Fox, Frances Margaret. *Flowers and Their Travels.* New York: Bobbs-Merrill Company, 1936.

Gibson, William H. *Sharp Eyes: A Rambler's Calender of Fifty-Two Weeks Among Insects, Birds and Flowers.* Massachusetts: Harvard Press, 1904.

Heiser, Charles B. *Weeds in My Garden: Observations on Some Misunderstood Plants.* Portland, Oregon: Timber Press, 2003.

Hopman, Ellen Evert. *Walking the World in Wonder: A Children's Herbal.* Rochester, Vermont: Healing Arts Press, 2003.

Hooker, Worthington M.D. *Plants: The Child's Book of Nature: For The Use of Families and Schools.* New York: Harper and Brothers, 1887.

Kavasch, Barrie. *Native Harvests: Recipes and Botanicals of the American Indian.* New York: Vintage Books, 1979.

Keeler Harriet L. *Our Native Trees: And How to Identify Them.* New York: Charles Scribner's Sons, 1900.

Kohl, George Johann. *Kitchi-Gami: Life Among the Lake Superior Ojibway.* Minnesota: Minnesota Historical Society Press, 1985.

Leopold, Aldo. A. *Sand County Almanac.* New York: Oxford University Press, 2001.

Leslie, Clare Walker and Roth Charles E. *Nature Journaling: Learning to Observe and Connect With the World Around You.* Pownal, Vermont: Storey Communications, 1996.

Lovejoy, Sharon. *Hollyhock Days.* Colorado: Interweave Press, 1991.

Louv, Richard. *Last Child In The Woods: Saving Our Children From Nature Deficit Disorder.* North Carolina: Algonquin Books of Chapel Hill, 2005.

Medlin, Julie J. *Michigan Lichens.* Bloomfield Hills, Michigan: Cranbrook Institute of Science, 1996.

Moerman, Daniel E. *Native American Ethnobotany.* Portland, Oregon: Timber Press, 2006.

Morely, Margaret W. *Little Wanderers.* Boston: Gin & Company, 1900.

Nichols, Robert, Jr. *Birds Of Algonquin Legend.* Ann Arbor, Michigan: The University of Michigan Press, 1995.

Peattie, Donald. *A Natural History of Western Trees.* Boston: Houghton Mifflin Company, 2007.

Pierre Yvette La. *Native American Rock Art: Messages from the Past.* Charlottesville, Virginia: Thomasson-Grant, 1994.

Riotte, Louise. *Sleeping With a Sunflower: A Treasury of Old-Time Gardening Lore.* Pownal, Vermont: Storey Communications, 1986.

Sanders, Jack. *The Secrets of Wildflowers: a Delightful Feast of Little-Known Facts, Folklore and History.* Connecticut: Globe Pequot Press, 2003.

Sisson, Edith. *Nature with Children of All Age: Activities & Adventures for Exploring, Learning, and Enjoying The World Around Us.* New York: Prentice Hall Press, 1982.

Skinner, Charles M. *Myths and Legends of Flowers, Trees, Fruits and Plants: In all Ages and In All Climes.* London: J.B Lippincott Company, 1911.

Starcher, Allison Mia. *Good Bugs For Your Garden.* North Carolina: Algonquin Books of Chapel Hill, 1995.

Stokes, Donald and Lillian. *A Guide to Observing Bird Behavior: Volume One and Two.* Boston: Little, Brown and Company, 1978 and 1983.

Stokes, Donald and Lillian. *A Guide to Nature in Winter.* Boston: Little, Brown and Company, 1976.

Stokes, Donald and Lillian. *A Guide to Enjoying Wildflowers.* Boston: Little, Brown and Company, 1985.

Stokes, Donald & Lillian. *A Guide to Observing Insect Lives.* Boston: Little, Brown and Company, 1983.

Stoutland, Allison. *Take a Deep Breath: Little Lessons From Flowers for a Happier World.* Lansing,Michigan: Inch By Inch Publications, LLC, 2002.

Tallamy, Douglas W. *Bringing Nature Home.* Portland, Oregon: Timber Press, 2009.

Vinal, William Gould. *Nature Recreation.* New York: Dover Publications, 1963.

Wilber, C. Keith M.D. *The Illustrated Living History Series: The Woodland Indians.* Connecticut: Globe Pequot Press, 1995.

Wilkins, Malcolm. *Plantwatching: How Plants Remember, Tell Time, Form Relationships and More.* New York: Facts on File, 1988.

Articles

Mann, Charles C. September, 2008. Our good earth, the future rests in the soil beneath our feet, understanding the importance of soil conservation. *National Geographic.* 214 (3):80-106

Klinkenborg, Verlyn. November, 2008. Our vanishing night, why we need darkness and the simple solutions to ending light pollution. *National Geographic.* 214(5):102-12

Weber, Larry. 2004. Teaching with the seasons, pp. 2-5. Grant, Tim and Littlejohn, Gail (ed) *Teaching Green: The Middle Years.* New Society Publishers, 2004.

Web sites

Solomon H. Katz. Contemporary issues. *Encyclopedia of Food & Culture.* Ed. Vol. 1. Gale Cengage, 2003. eNotes.com. 2006. 13 Nov, 2010. http://www.enotes.com/food-encyclopedia/contemporary-issues.

Rollins, Alice F. *Industrial Work for Public Schools.* Chicago, New York: Rand, McNally and Company, 1860. http://www.archive.org/details/cu31924013381722

Citizen Science

Plant Watch, www.plantwatch.ca

Project Budburst, www.windows.ucar.edu/citizen-science/budburst/collaboorators.php

Earth Trek Project, www.goearthtrek.com

Citizen Science Guide/Cornell Lab of Ornithology and Audubon Society, www.birdsource.org/gbbc/get-involved

Nature's Calendar, www.naturescalendar.org.uk

Phenology

Phenology, Nature's Cycles of Life, www.sws-wis.com/lifecycles

Applied Phenology and Gardening, www.wihort.edu/landscape/phenology.htm

National Sustainable Agriculture Information Service, www.attra.org/atta-pub/phenology.htm

Backyard Nature, www.backyardnature.net

Wisconsin Phenology Society, www.naturenet.com/alnc/wps

Credits

Introduction

Carson, Rachel. *The Sense of Wonder,* Harper and Row, 1956, p. 88-89. Copyright © By Rachel L. Carson. Reprinted by permission of Frances Collin, Trustee.

American Indian Spring Legend

Rich, Edwin Gile. Why the robin brings spring. Adapted from: *Why So Stories.* Boston: Small, Maynard and Company, 1920, pp. 33-38.

Spring

Matson, Ruth. There comes a day. New York: *The Home Gardener.* March, vol. 11, no. 3, 1948, p. 33.

March

Nygaard, Anita. *Earthclock: A Narrative Calendar of Nature's Seasons.* Harrisburg,

Pennsylvania: Stackpole Books, 1976, p. 206.

Artz, Jerie. Watercolor based Beaufort-Wind Scale.

April

Gardener's Supply Company, *Build Your Own Rain Garden, Perennial Garden Design Sheet #1*. Burlington, Vermont, www.gardens

Purslane, black and white drawing, USDA-NRCS PLANTS Database/Britton, N.L., and A. Brown. 1913. An Illustrated Flora of the Northern United States, Canada and the British Possessions. 3 vols. Charles Scribner's Sons, New York. Vol. 2: 40.

Gail, Peter A. *The Dandelion Celebration,* Cleveland, Ohio: Goosefoot Acres Press, 1994. Dandelion flower cookies, p.111, Dandelion pizza, p.81.

Gail, Peter A. *Volunteer Vegetable Sampler: Recipes for Backyard Weeds*, Cleveland, Ohio: Goosefoot Acres Press, 1991. Baked nettles and potatoes, p.33. Clover soup p.43. Bacon-coated purslane, p.41. (For more information: P.O. Box 446, Valley City, Ohio 44280-0446. Web site: www.dandyblend.com or Phone 800-697-4858)

May

Weber, Nelva M. Homemade timetable for seed sowing. New York: *The Home Garden,* April, 1947, vol.9 no.1, p.18. This list was adapted with permission from Nelva Weber's daughter Diana Sammataro.

Carson Rachel. *The Sense of Wonder*, Harper and Row, 1956. p.49.
Copyright © By Rachel L. Carson. Reprinted by permission of Frances Collin, Trustee.

American Indian Summer Legend

Larned, W. T. L. How summer came. *American Indian Fairy Tales.* Chicago, Illinois P. F. Volland Company, 1921, Story #6

June

Artz, Jerie. Strawberry photo

Ewing, Nancy. Butterfly gardening and making a butterfly puddle

Milkweed black and white drawing, SDA-NRCS PLANTS Database/Britton, N.L., and A. Brown. 1913. *An illustrated flora of the northern United States, Canada and the British Possessions. 3 vols.* Charles Scribner's Sons, New York. Vol. 3: 28.

July

Hansen, Michael. August, 2009. Michigan pollinators. *The Michigan Landscape* p.25.

Xerces Society for Invertebrate Conservation. *Nests for Native Bees Fact Sheet* and *Upper Midwest Plants for Native Bees Fact Sheet*

August

Belsinger, Susan and Wilcox, Tina Marie. Jewel weed vinegar. *the creative herbal home.* Brookeville, Maryland: herbspirit, 2007, p.111.

Caduto, Michael J. and Bruchac Joseph. *Native American Gardening: Stories, Projects and Recipes for Families.* Colorado: Fulcrum Publishing, 1996, p.5.

Liriodendron tulipifera L. Tuliptree/photo by William S. Justice @USDA-NRCS PLANTS Database, courtesy of Smithsonian Institution. http://plants.usda.gov

American Indian Autumn Legend

Larned, W.T.L. Shin-Ge-Bis Fools the North Wind. *American Indian Fairy Tales.* Chicago, Illinois: P. F. Volland Company, 1921, Story #2 (With one exception, all the tales in this book are adapted from the legends collected by Henry J. Schoolcraft)

September

French, Coleen. Plantain salve recipe and plantain plant photo

Wittgen, Gudren. Bird silhouette cutting

Byrd, Deborah. 2002. Harvest moon (online). EarthSky. A Clear Voice for Science (cited 22 September 2010) Available from the World Wide Web Earthsky.org

Densmore, Francis. *Chippewa Customs.* Minnesota: Minnesota Historical Society, 1979, p.127.

French, Coleen. Squash stem whistle photo

October

Beals, Katherine M. The goldenrod. *Flower Lore and Legend.* New York: Henry Holt Company, 1917. pp. 215-218.

November

Belsinger, Susan. Rock turtle photo

Betz, Demian. Beaver photo

Ruoff, Henry. *The Century Book of Facts.* Springfield, Massachusetts: King-Richard Company 1902, p.331.

Field horsetail drawing, USDA-NRCS PLANTS Database/USDA NRCS. Wetland flora: Field office illustrated guide to plant species. USDA Natural Resources Conservation Service.

Witch hazel Tree photo, W.D. Brush @ USDA-NRCS PLANTS Database

American Indian Winter Legend

Jameson, Anna. Allegory of winter and summer. O-sha-gush-ko-da-na-qua the eldest daughter of Chippewa chief Waub-Ojeeg, recounted this story to Anna Jameson during her visit to Ontario Canada in 1836. *Winter Studies and Summer Rambles In Canada.* Canada: McClelland & Steward, 1923, p.366-368.

December

Hottes, Alfred Carl. *Garden Facts and Fancies.* New York: Dodd, Mead & Company, 1949, p.151.

Eastern Red Cedar photo courtesy USDA, NRCS. 2011. The PLANTS Database http://plants.usda.gov, 22 March 2011. National Plant Data Center, Baton Rouge, LA 70874-4490 USA.672

January

Nygaard, Anita. *Earthclock: A Narrative Calendar of Nature's Seasons.* Harrisburg, Pennsylvania: Stackpole Books, 1976, p.190.

Heidorn, Keith C. PhD. The January thaw (On line) <u>The Weather Doctor Almanac</u> (cited 15, January 2002) Available from World Wide Web (<u>http://www.island-net.com/~see/weather/doctor.htm.</u>)

Comstock, Botsford Anna. *The Handbook of Nature Study.* Ithaca, New York: Cornell University Press, 1955, p.672.

February

Forestfarm. 2010 Winter catalog. Try making paper yourself. p.14.

Forestfarm 990 Tetherow Rd. Williams, OR 97544-9599 <u>www.forestfarm.com</u>

Index